AF787374

Gasification - Current Technologies and Future Prospect

Edited by Yongseung Yun

Published in London, United Kingdom

Gasification - Current Technologies and Future Prospect
http://dx.doi.org/10.5772/intechopen.1005984
Edited by Yongseung Yun

Contributors
Alexei Mikhailovich Essiptchouk, Ali Can Sivri, Asmare Tezera Admase, Daniel Martín Márquez López, Ejigayehu Desalegn Asrade, Ektaí López Ángeles, Eshetu Getahun, Fredy Josealdo Castillo Plata, Gerardo de los Santos Camas, Jin Wook Lee, José Luz Silveira, Karittha Im-orb, Kunlanan Wiranarongkorn, Regina Franciélle Silva Paulino, Seok Woo Chung, Sergio O. Flores Valle, Seung Jong Lee, Tesfa Nega Gesese, Yongseung Yun, Yongseung Yun

© The Editor(s) and the Author(s) 2025

The rights of the editor(s) and the author(s) have been asserted in accordance with the Copyright, Designs and Patents Act 1988. All rights to the book as a whole are reserved by INTECHOPEN LIMITED. The book as a whole (compilation) cannot be reproduced, distributed or used for commercial or non-commercial purposes without INTECHOPEN LIMITED's written permission. Enquiries concerning the use of the book should be directed to INTECHOPEN LIMITED rights and permissions department (permissions@intechopen.com).

Violations are liable to prosecution under the governing Copyright Law.

Individual chapters of this publication are distributed under the terms of the Creative Commons Attribution 4.0 License which permits commercial use, distribution and reproduction of the individual chapters, provided the original author(s) and source publication are appropriately acknowledged. If so indicated, certain images may not be included under the Creative Commons license. In such cases users will need to obtain permission from the license holder to reproduce the material. More details and guidelines concerning content reuse and adaptation can be found at http://www.intechopen.com/copyright-policy.html.

Notice

Statements and opinions expressed in the chapters are these of the individual contributors and not necessarily those of the editors or publisher. No responsibility is accepted for the accuracy of information contained in the published chapters. The publisher assumes no responsibility for any damage or injury to persons or property arising out of the use of any materials, instructions, methods or ideas contained in the book.

First published in London, United Kingdom, 2025 by IntechOpen
IntechOpen is the global imprint of INTECHOPEN LIMITED, registered in England and Wales, registration number: 11086078, 167-169 Great Portland Street, London, W1W 5PF, United Kingdom

For EU product safety concerns: IN TECH d.o.o., Prolaz Marije Krucifikse Kozulić 3, 51000 Rijeka, Croatia, info@intechopen.com or visit our website at intechopen.com.

British Library Cataloguing-in-Publication Data
A catalogue record for this book is available from the British Library

Gasification - Current Technologies and Future Prospect
Edited by Yongseung Yun
p. cm.
Print ISBN 978-1-83634-503-9
Online ISBN 978-1-83634-502-2
eBook (PDF) ISBN 978-1-83634-504-6

If disposing of this product, please recycle the paper responsibly.

IntechOpen
intechopen.com

Built by scientists, for scientists

Explore all IntechOpen books

Meet the editor

Dr. Yongseung Yun majored in chemical engineering and received his Ph.D. from the University of Utah in 1990. He also holds a Master's degree from KAIST (Korea Advanced Institute of Science & Technology) in Korea, obtained in 1981, and a BSc degree from Yonsei University in Korea, earned in 1979. He had worked as a vice president and is currently a distinguished researcher at the Institute for Advanced Engineering in Korea. At KAIST from 1981 to 1984, he conducted research on coal-oil agglomeration and fluidized bed coal combustion. For his Ph.D., he worked on the low-temperature oxidation of coal using air and developed a TG/MS instrument for coal analysis. At Brown University in 1991-1992, he researched the pre-treatment methods to enhance coal liquefaction yield. From 1993 to the present, he has focused on process developments to produce clean synthetic gas from coal, pet coke, and waste, including biomass. Currently, he heads the blue hydrogen project, which includes 30 tons/day petcoke gasification, syngas cleaning, and PSA purification for 2 tons/day of high-purity hydrogen production. Between 2013 and 2018, he served as the president of the Korea Association of Waste to Energy (KAWET) and as Chief Editor of the Korean Industrial Chemistry News for the Korean Society of Industrial and Engineering Chemistry (KSIEC) for eight years. He currently acts as vice president of the Korea DME Association.

Contents

Preface

Gasification technology has been in use for almost 100 years and has undergone extensive upgrades over the last four decades, making it applicable to high-pressure/higher efficiency plants and meeting increasingly stringent environmental requirements. It is in a position to be challenged by recent circumstances of climate-concerned demand and of sustainability in employed feedstock. Basically, fossil-fuel-based syngas is being challenged by renewable syngas. Renewable syngas production is currently in the demonstration stage and is proving to be a viable, commercial-ready industry. Most competition arises in application areas for renewable green hydrogen and for renewable biogas. Fisher-Tropsch synthesis, the most traditionally important application area of gasification, is being challenged by a process in which renewable hydrogen from electrolysis powered by renewable electricity and captured CO_2 is reacted to produce sustainable hydrocarbon fuels, such as sustainable aviation fuel (SAF). It should be emphasized that there are many greener and more socially anticipated options for end-products from gasification nowadays, such as SAF, e-methanol, and renewable diesel. Syngas, a mixture of carbon monoxide and hydrogen, can be produced through the reaction of renewable biogas (rich in methane) via steam-methane reforming and also through CO_2-methane dry reforming, which utilizes greenhouse-gas CO_2, or even through the electrochemical reduction of CO_2 and water. All renewable syngas routes are still quite expensive at the current level; however, societal demand dictates that they will eventually overcome the cost barrier.

Gasification offers several advantages in converting low-grade organic materials into syngas, characterized by lower pollutant emissions and more cost-effective CO_2 capture potential. However, it also presents several drawbacks, including process complexity due to the nature of syngas and the associated high instrumentation requirements. Considering the many advantages, there can be sufficient room even for fossil-fuel syngas, provided proper upgrading and the incorporation of recently developed technologies are employed. Gasification technologies must evolve by integrating recent advancements in catalysts, adsorbents, separation, sensors, remote control, and artificial intelligence (AI) tools. Moreover, gasification can effectively utilize advancements in syngas-related equipment, such as gas turbines, fuel cells, and membranes, to overcome economic limitations.

A niche market for gasification exists in many under-utilized feedstocks such as biomass and wastes, because a large portion of them are renewable in nature. Recent developments in gasification markets utilizing biomass and waste provide a significant impetus for future gasification technologies that can evolve to be carbon-neutral and environmentally benign.

In the recent gasification trend, renewable syngas production methods are gradually replacing conventional fossil-fuel-based methods at a slow but steady pace. This trend is inevitable and beneficial to the industry and to people. Conventional gasification methods still possess many beneficial features that can adapt to the CO_2-conscious world, aided by adjacent technologies in addressing CO_2 issues and securing sustainability with biomass and waste feeds.

This book discusses the application areas and fundamental issues encountered when gasification technologies are utilized, with further illustration regarding the future direction of gasification. Based on the realistic evaluation of current gasification technology and analysis of acute drawbacks of gasification with feasible future directions, this book provides a glimpse that hopefully inspires future upgrading of gasification technologies.

The book consists of seven chapters, divided into three sections. The first section comprises a single chapter for a general discussion on the current status and future direction of gasification technologies. In particular, Chapter 1 covers the multifaceted aspects of gasification technologies, starting from the fundamental benefits/inherent issues in gasification and its application areas, and finally leading to future direction points.

Two chapters in the second section deal with hydrogen production by gasification: one for blue hydrogen production using pet coke and another titled 'Hydrogen Production via Sustainable Gasification'. Blue hydrogen is a viable solution to mitigate the CO_2 issue when fossil fuels are used as feedstock for gasification. Although the green hydrogen produced by water electrolysis, starting from renewable electricity from wind and solar power, should be the ultimate goal, blue hydrogen can fill the gap until green hydrogen achieves the acceptable cost target. Chapter 2 describes the actual operational data in pilot and demonstration plants on the hydrogen production using petroleum coke. A demonstration plant for producing two tons of hydrogen per day consisted of a 30-ton-per-day gasifier, desulfurization unit, water-gas shift reaction unit, and pressure swing adsorption unit. Chapter 3 presents the recent novel hydrogen production technologies and compares the CO_2 amount produced from different process configurations.

The third section, comprising four chapters, explores the application of technologies utilizing the important, yet underutilized, feedstock — biomass and biomedical waste — that can demonstrate the efficacy of gasification technology. Chapter 4 provides introductory information on the effects of operational variables in biomass gasification, as well as the challenges involved in maximizing the potential of biomass feedstock. Chapter 5 describes the catalytic gasification of biomass, providing information on the catalysts employed in hydrogen production. Chapter 6 describes the fundamental characteristics of a bubbling fluidized-bed pyrolysis reactor using biomass feed, with in-depth monitoring of mixing quality at different bed aspect ratios. Chapter 7 presents a unique case study in which energy/cost analysis results are discussed when applying plasma gasification to biomedical waste in Brazil.

I would like to thank all the authors who shared their valuable experiences and data on gasification for this book. Additionally, I would like to express my sincere thanks to Publishing Process Manager Ms. Marica Novakovic at IntechOpen, who worked diligently throughout the reviewing/editing process over the past ten months.

Yongseung Yun
Institute for Advanced Engineering,
Yongin, Republic of Korea

Section 1

Current Status and Future Prospect

Chapter 1

Current Status and Future Direction of Gasification Technologies

Yongseung Yun

Abstract

Gasification technology has been around for almost 100 years and is in a position to be challenged by recent circumstances of climate-concerned demand and of sustainability in employed feedstock. Gasification has many advantages in converting low-grade organic materials into syngas with inherent features of lower pollutants and CO_2 capturing potential as well as many drawbacks in terms of process complexity due to the nature of syngas and related high instrumentation. Application areas and fundamental issues encountered when gasification technologies are utilized are discussed, with further illustration regarding the future direction of gasification. Gasification can best use the achievements in syngas-related equipment like gas turbines, fuel cells, and membranes to overcome the economic limitations, and it can use the recent developments in catalysts, adsorbents, and separation areas. Recent gasification markets in using biomass and wastes provide a significant impetus for future gasification technologies that can evolve to be carbon-neutral and environmentally benign.

Keywords: gasification, syngas, hydrogen, CO_2, future direction

1. Introduction

Gasification has been known for 100 years and had been applied to coal and petroleum residues at commercial scale. Most commercial success came from indirect coal liquefaction in South Africa. Sasol plants that employed fixed-bed type gasifiers flourished in the 1970s during the oil crisis and in the 2000s when oil prices shot up to US$140/barrel. After crude oil prices stabilized below US$90/bbl and as more climate concerns on CO_2 emerged, gasification technologies have shifted gear toward applications to high-efficiency, carbon-neutral, low-grade feedstock and to high-value products.

The ultimate goal of gasification should be the conversion of each component in organic feedstock into high-value end products. Most visible organic feed involves coal, natural gas, petroleum residues, biomass, and wastes. Most typical high-value end products comprise electricity, hydrogen, synthetic diesel, synthetic natural gas, methanol, and di-methyl ether (DME).

As the proverb goes, the devil is in the details for the gasification case. Gasification itself looks so handy and well proven for use in converting any organic materials into useful gaseous products while minimizing environmental agonies. But the reality is far from easiness and all-proven use. Rather, it is very dirty and harmful in current processes. Gasification technology was not matured enough to deal with several hundreds of tons per day scale for waste feedstocks, which have uneven and irregular characteristics.

When technical experts are involved in large-scale commercial plants that have invested enough budget to cover all the details, performance in gasification processes can be satisfactory. However, if not enough qualified fabricators or operators are engaged, performance can be in disaster levels as demonstrated in many earlier commercial experiences.

Gasification projects of scales up to about 20 tons/day (approximately 1 ton/hour scale) in the simple forms of reactors have a good successful history. Also, scales of 2500–3000 tons/day such as integrated gasification combined cycle (IGCC) plants have demonstrated a good technical success record. In cases of scale that goes above 200 tons/day but not as big as IGCC, there are only limited success stories in using fixed-bed reactor systems as in South African Sasol plants using gasifiers in parallel, each with a scale of 1300 tons/day. Although many ambitious gasification projects were conducted during the 2010s using plasma gasification technologies for wastes in England and in Southeast Asian countries, all have failed to reach commercial operations yet due to unsatisfactory performance in gasification.

It should be noted that explicit competition exists in gasification applications. In fact, there are many greener and more socially anticipated options for end products from gasification nowadays, such as sustainable aviation fuel (SAF), e-methanol, and renewable diesel. Gasification technologies have to evolve by fusing with recent developments in catalysts, adsorbents, separation, turbines, and artificial intelligence (AI) tools.

Most fundamental decisions will depend upon the deliverability of bankable project with extremely low levels of technical and social risk. Social risk definitely involves net CO_2 emissions and people's fear of harmful aspects of syngas.

Whole green electricity from solar and wind must be the ideal final destination of energy, but they are unfortunately still expensive and capital intensive in most cases. Feedstock of organic nature including fossil fuels will remain quite a long time as an energy source, at a declining trend. Gasification is well viewed as an interim route that can utilize biomass and wastes whenever the raw feedstock is combustible and of organic nature.

Biomass has been regarded as a carbon-neutral fuel. But direct combustion (burning) can cause significant CO_2 emission. Some research argues that biomass can release more carbon dioxide per unit of energy than coal or gas without employing proper systems. Regardless, the European Union declared biomass as a carbon-neutral energy source in 2009, which was a starting point of wide usage of biomass for mitigating CO_2 emission. Biomass is one important large-scale, non-intermittent, low-carbon solution that can provide cost-effective electricity and hydrogen. Gasification can provide a more efficient method than direct combustion in using biomass.

When considering the distributed energy infrastructure, municipal solid wastes (MSW) and organic wastes like food waste can come first since they are in the vicinity of energy-requiring areas.

Key challenges in gasification lie in aspects of CO_2 emission, safety issues, technical completeness, scalability, and long-term competition with e-fuels and so on.

Examples of large-scale commercial application can be found in IGCC areas. IGCC plants using coal in Japan and Korea as well as plants in Saudi Arabia and in India using petroleum coke are typical examples.

A niche market exists in many under-utilized feedstocks and in many developing countries where abundant biomass and waste resources are available. Biomass is abundant in most developing countries, but utilizing it as an energy source should involve careful selection of technology. In addition, many developed countries prohibit direct landfill of MSW and thus require converting all solid combustibles to gas or liquid forms. Making plastic circularity through gasification starting from sorted waste plastics to the end product of synfuel or hydrogen is currently a hot area to compete for commercial application.

It will be worthwhile to review the current status of gasification technology itself, its acceptance in society that asks more greener and easier to use, and to anticipate what will be the future direction of gasification for the coming green/carbon-neutral energy era.

Many excellent books are available about gasification technology. By this time it should be worthwhile to reconcile both sides: the technical aspect as well as the socially cherished carbon-conscious trend. Based on the realistic evaluation of gasification technology and focusing on acute drawbacks of gasification by comparing it with newly emerging carbon-free competing technologies, this chapter provides a glimpse that hopefully inspires future upgrading of gasification technologies.

2. Fundamental benefits of gasification

2.1 Distinctive advantages of gasification

Key technical pathways in breaking or converting organic feedstock involve thermal methods or biological methods. Thermal methods are relatively fast reacting and easy to scale up, while biological methods typically follow a slower step under limited scale-up capability as one unit. Biological gasification in which biomethane is normally retrieved can yield a significant impact on future gasification applications, but the thermal approach will be the main topic in this chapter.

Depending on the oxidant supply amount, the thermal approach divides into pyrolysis, gasification, and combustion. Distinctive aspects of gasification rely on the conversion form of N and S components in organic materials. All terrestrial organics are composed of C, H, O, N, and S components. A fundamental objective of gasification is in converting C and H components in organic materials into useful forms of typically aliphatic C-H chains like gasoline or into pure carbon monoxide and hydrogen. Nitrogen and sulfur compounds come out as pollutants if oxidant is provided excessively as in combustion. As shown in **Table 1**, combustion produces NOx and SOx with enough oxygen supply. Gasification supplies limited oxygen. Then available oxygen is consumed quickly by the very fast carbon-oxygen reaction at the high temperature region before N and S components are doing anything with oxygen. N and S then only have an option of reacting with only the available or remaining H component, and they finally end up in HN_3 and H_2S forms. These forms are easier to capture than SOx and NOx. NH_3 (ammonia) is used as a feed for fertilizer, and H_2S can transform to sulfuric acid or elemental sulfur through the well-established Claus process in the petroleum industry. Therefore, N and S components can become sellable products instead of pollutants that require another cost burden in post-removal procedure.

Item	Pyrolysis	Gasification	Combustion
Temperature (°C)	300–850	600–1600	850–1450
Pressure (bar)	1	1–45	1
Supply gas	Inert gas (N_2)	O_2, air, steam	Air, O_2-enriched gas
Oxidant excess ratio	0	<1	>1
Main reaction	Organics decomposition	$C + H_2O = CO + H_2$ $C + CO_2 = 2CO$	$C + O_2 = CO_2$ $H_2 + 1/2O_2 = H_2O$
Products			
• Gas phase	Hydrocarbon, H_2, CO	H_2, CO, CO_2, CH_4	CO_2
• Solid phase	Ash, tar, char	Slag	Ash
• Liquid phase	Pyrolysis oil, water	Water	Water
Main pollutants	HCl, NH_3, CO, SOx, NOx, H_2S	H_2S, HCN, COS, NH_3, CO	SOx, NOx
Main purpose	Oil replacing crude oil	Fuel gas, electricity/steam, chemicals	Electricity, steam

Table 1.
Key differences among pyrolysis, gasification, and combustion processes.

Ash component in feeds can convert into slag under high temperatures which are above the ash melting temperature in the reactor types of entrained-bed and plasma. During the liquefied period of melted ash, harmful inorganic components (Pb, Cr, Cd, As, Ni, etc.) would mixed with main ash components (Al_2O_3, SiO_2). When the melted ash exits the reactor and positions under lower temperatures, harmful components are trapped inside the rock-like slag structure during solidification. Then these harmful components cannot escape to the environment under normal atmospheric situations like under rain. Most countries classify the safety of produced slag by leachate concentrations of heavy metals. Well-treated slag under high temperatures will surely suffice for the regulation limits.

Table 2 illustrates the extraordinary environmental performance by gasification-based IGCC power plants that are currently in operation [1]. Since syngas is utilized for electricity generation in IGCC, environmental regulations that apply to natural gas power plant are compared. Natural gas is touted as the most clean fossil fuel replacing other fossil fuels.

In Korea's case, allowed SOx and NOx concentrations are 20 ppm and particulates regulation value is 10 mg/Nm^3. Environmental performance in IGCC power plants clearly demonstrates the environmental excellence of gasification, which shows that gasification can provide a cleaner method than natural gas if it can garner enough investment.

Fine particulates (dust in regulation term) are far lower than for a liquefied natural gas (LNG) power plant. This is mainly due to the requirement for gas turbines that are used in combined cycles like in IGCC. To protect the turbine blades that rotate at >3500 rpm, any particulates that can damage the blade have to be removed. For this purpose, IGCC typically applies metal filters instead of normal liquid-type scrubbers. This increases capital/operating costs but will surely remove almost all particulates, which results in low values as in **Table 2**.

Item	Nakoso coal IGCC plant, Japan	Osaki coal IGCC plant, Japan	GreenGen coal IGCC plant, China [2]	Taean coal IGCC plant, Korea	LNG power plant, environmental regulation level, Korea (2018–)
Electricity generating capacity (MW, gross)	250	166	265	300	—
SOx (ppm)	1.0	8	1.0	<0.5	20
NOx (ppm)	3.4	5	13.5	<3–4*	20
Dust (mg/Nm^3)	<0.1	3	0.29	<0.1	10

**Measured as 3–4 ppm, down from earlier 24 ppm level after installing SCR in 2018.*

Table 2.
Actual pollutants emission amount in currently operating commercial coal IGCC plants and comparison with environmental regulation level of natural gas power plant in Korea [1].

2.2 Harnessing benefits of gasification

The gasification process is a good method in converting low-grade combustible materials into higher-value energy source materials. As the wording specifies, gasification is suitable for generating a gaseous energy source, especially in gas-based and in electricity-based society. Syngas itself and synthetic natural gas have helped ever increasing clean gas demand, and syngas has used in IGCC to generate relatively clean electricity except for CO_2 issues.

The beauty of gasification lies in its ability to convert all combustibles to gaseous form in a short time. Gasification needs to heat to the reaction temperature to proceed. The cheapest option for heating is utilizing some portion of raw feedstock instead of another outside heating source. Most gasification processes rely on the process concept that uses a certain amount of cheap, low-grade raw feedstock for increasing the temperature to the gasifying range. Gasification utilizes some part of feedstock such as coal, petroleum residues, biomass, and wastes, which are the cheapest energy source at the plant site, for maintaining high temperatures. It should definitely be cheaper than using electricity or natural gas. Outside heating sources can be applied to minimize unwanted CO_2, but they entail a cost for that. The gasification method of $C + H_2O \rightarrow CO + H_2$ will be the best way for producing syngas ($CO + H_2$) without CO_2 emission, but it requires outside energy to maintain the reaction temperature of typically above 900°C, in many cases above 1200°C. There are also other gasification routes like CO_2 gasification or heating by plasma or microwave. All these outside heating methods, rather than direct utilization of some portion of the plant feed that exists at the plant site, will be expensive options.

Cold gas efficiency (CGE) is defined as the ratio of heating value in produced syngas divided by heating value that raw feedstock possesses. Most of the energy deducted from 100% by cold gas efficiency will be used for increasing the temperature inside the gasifier. Heat loss from the reactor and the energy entailed with exiting hot syngas and fines from the gasifier will be another portion in lowering cold gas efficiency. Large-scale gasifiers yield higher CGE. Commercial coal gasifiers exhibit CGE of around 80%. In waste gasification at smaller scales, CGE can range as low as less than 50%.

Why does the gasification reaction require operation at high temperatures? As discussed previously, typical methods of breaking organic materials are pyrolysis, liquefaction, combustion, and gasification. When oil prices were high, pyrolysis and liquefaction methods were widely utilized to replace expensive crude oil and oil products. When the oil price is less than US$60–80/barrel, economics are not sufficient to support synthetic fuels from gasification. Another important factor in lowering the technology demand of pyrolysis and liquefaction types is the strict compliance mandate for environmental regulations that put a harsh burden for small or moderate sizes of plant. Principles of pyrolysis and liquefaction rely upon extracting only the valuable oily substances while retaining solid and tarry interim products. If these interim products can transform as one that can be sold, it can overcome the cost barrier. But most of these interim products rather demand additional treatment facilities with cost. Of course, pyrolysis is well suited in producing oil products that can replace petroleum oil by using waste plastics of relatively clean status.

In the case of liquefaction, the requirement of high water consumption and high CO_2 emission rate hampers any economic application except a large-scale plant of several thousand tons per day. An earlier US Department of Energy (DOE) report [3] explained that transportation fuels made from coal by liquefaction resulted in 7–10 times higher CO_2 production than those from converting crude oil.

Therefore, instead of worrying about all intermediate byproducts and all treatments to each product sequence, a simple method can be conceived in breaking all organic components into very simple forms of gas compounds like CO and H_2 under very high temperatures. Typically organic compounds can quickly break at temperatures above 1200°C. To make sure of the fast reaction in bond-breaking and to make sure even inorganics can melt, temperatures above 1400°C are employed in gasification types of entrained-bed and plasma torch. In fluidized-bed type gasification, a typical temperature range is 700–900°C, where ash will remain at a similar size as internal heat-carrying medium like sand. Ash in a fluidized bed should work as a part of the fluidizing solid that keeps the internal temperature during the process uniform. Therefore, ash should not convert to slag, which will have a higher density that hampers fluidization.

In fixed-bed gasifiers, there are two technology options of ash-nonmelting and ash-melting. In fixed-bed application using biomass, the usual choice is the nonmelting type, since biomass does not contain much inorganic amount compared to coal or MSW. The nonmelting type is obviously preferred if possible, which consumes less energy by relatively lower temperature operation. Another important factor determining gasification temperature is the nature of the gasifier inner wall, whether it is a refractory or a wall of steam-generating metal tube coils. Refractory wall is cheaper and easy to install, but service life drops quickly under higher operating temperatures above 1600°C by intermittent hot spots.

3. Current application areas of gasification

Gasification has been applied to many energy-related areas. The most prominent one is synthetic fuels in the 1930s–1940s and during the 1980s of the oil crisis period. During the 1980s when oil prices once again shot up, coal or petroleum residue based integrated gasification power plants were envisioned and materialized. After climate change concerns arise, application of gasification to biomass and wastes becomes a main topic.

3.1 Clean electricity

This is the most common and successful application of gasification. During the 1970s, a new concept for coal-fired power plants was developed to adopt a combined cycle that was widely used in natural gas-fired power plants. Syngas from coal gasification can meet the criteria, and IGCC technology materialized. Electricity from IGCC can be called relatively clean compared to conventional combustion-based steam power plants in terms of lower pollutant emission and higher efficiency. **Table 2** in an earlier section illustrates the current commercial IGCC plants. **Figure 1** shows the plant view of the Taean 300 MW IGCC power plant in Korea [4].

Electricity generation through gasification using locally available biomass or wastes is a feasible option in rural or remote areas detached from a centralized grid. Internal combustion engines are currently used, which has a lower efficiency than large-scale gas turbines and combined cycles.

3.2 Synthetic fuels and chemicals

The Fisher-Tropsch (FT) process that first developed around the 1920s in Germany has been widely applied. Nowadays a small and modular type of FT system is being tried to syngas from biomass and wastes. The most vivid example of synthetic fuels is South Africa's Sasol plant using coal in more than 40 fixed-bed gasifiers in parallel. A synthetic natural gas (SNG) plant in North Dakota, United States, has been operated from 1984.

Many petroleum companies are employing medium-size gasification plants for the purpose of producing pure CO and H_2. Syngas is a valuable feed for manufacturing acetic anhydride, acetic acid, methyl acetate, and methanol. Recently an alcohol-to-SAF route using a gasification process with waste feed is in the commercial stage with an expected lower unit cost [5]. Low-carbon circular methanol from the waste feed is the key in this route.

Figure 1.
View of 300 MW$_e$ Taean coal IGCC plant in Korea, operating from 2018 [4].

3.3 Hydrogen

Syngas contains typically 30–60% CO, 25–30% H_2, less than 5% CH_4, 5–15% CO_2, water vapor, and sulfur/nitrogen compounds such as H_2S, COS, and NH_3. The CO portion in syngas is converted by a water-gas shift reaction to CO_2 and H_2, which increases H_2 concentration. In principle, hydrogen generation through syngas is an obvious route, but due to the cost involved in separation and in contaminants removal, actual economic application has been limited in commercial scale. Blue hydrogen that separates out CO_2 by pressure swing adsoprtion (PSA) or other means is becoming a good candidate in future energy mixes.

Direct separation of CO and H_2 from syngas by the pressure-swing method is typically operated at high pressure and shows a low gas recovery ratio. But separation of CO_2 and H_2, which are the main components after the water-gas shift reaction, is an easier route.

4. Inherent and fundamental issues in gasification

4.1 Monitoring gasification of fast reaction time and at high temperatures

This depends on the type of gasifier, whether it involves high temperature above 1200°C or not. A K-type thermocouple normally can endure up to 1200°C while maintaining a linear signal response to temperature variation.

At temperatures above 1400°C, the reaction rate is very fast so that the whole residence time in the gasifier can be a few seconds in the entrained-bed type and can be faster in the plasma type. In entrained-bed and plasma gasifiers which entail high temperature in a short time, R-type thermocouple is typically employed as the main temperature measuring tool. When a fast reaction of a few seconds residence time occurs under high temperature above 1400°C, then actual measurement by R-type thermocouples, which can stand up to 1600°C, becomes a challenge due to its shortened life expectancy. Actually, the internal gasifier itself becomes a black box where no direct measurement is feasible, which ends up with expensive monitoring devices that are needed to prevent any safety problems. In the actual case, without on-line temperature data during the few seconds of reaction time, responding to abnormal happening (e.g., internal refractory failure) becomes difficult and can pose a great safety risk. This situation causes the operation & maintenance (O&M) problem in the long term and induces lower plant availability in gasification operation.

In fluidized-bed and fixed-bed type gasifiers, the maximum temperatures are typically less than 900°C and 1200°C, respectively. Then, this problem is not severe since normal K-type thermocouples work well and the reactor residence time is in minutes, not in a few seconds.

4.2 Syngas cleaning

Syngas contains significant impurities that originated from nitrogen and sulfur compounds. When syngas is utilized for thermal purposes, conventional cleaning methods, which have been widely used in the petrochemical industry, can be applied. The lower-left part of the IGCC plant in **Figure 1** is the removal facility for impurity gases, which shows a very similar outlook of petrochemical plants.

Contaminant	Allowed contaminant level for application		
	Gas turbine	Methanol synthesis	Fisher-Tropsch products synthesis
Particulates	<30 mg/m^3	<0.02 mg/m^3	<0.01 μL/L
Alkali compounds	<0.02 μL/L	—	<0.01 μL/L
Sulfur compounds	<20 μL/L	<1 mg/m^3	<0.02 μL/L
Nitrogen compounds	<50 μL/L	<0.1 mg/m^3	<0.01 μL/L
Halides	<1 μL/L	<0.1 mg/m^3	<0.01–1 μL/L
Tars	—	<0.1 mg/m^3	

Note: 1 mg/m^3 = 1 ppb, 1 μL/L = 1 ppm.

Table 3.
Typically required contaminant levels for gas turbine and chemical synthesis [6].

When the final application of syngas is chemicals, the level of impurities removal goes down to the ppm and ppb level, as shown in **Table 3** [6]. It will be the best situation if the gas turbine and catalysts can endure the impurities to somewhat higher concentrations. But, unfortunately, the syngas specification should meet the gas turbine specification and the catalyst-allowed level in synthesis reactors. A typical method is first removing the bulk of the syngas to the ppm level and making it lower by adsorbent guard beds to the ppb level.

Since syngas holds a high concentration of CO, removing CO to the required level of catalyst-using applications that do not use CO as feed as well as to the environmentally allowed CO level imposes an operational and cost burden.

4.3 Inherent safety issues of syngas (CO and hydrogen)

Inherent toxicity of CO and flame/fire characteristics of hydrogen are the main ingredients to force the gasification plant to be complex and expensive. The CO concentration in syngas can reach 60%, which is 600,000 ppm. Even an air-fed fluidized gasifier using biomass exhibits at least several percent of CO even with dilution by nitrogen in air. Carbon monoxide is a well-known toxic gas. A CO concentration of 35 ppm requires oxygen masks, and 3000 ppm (0.3%) can lead to death in 30 minutes exposure. One accident in the author's lab in 2002 can be an example of the CO safety issue. One researcher who was in charge of on-line syngas analysis, which was an essential method to provide information for controlling the O_2 input amount and the syngas quality, found a loose 1/4″ Tygon tube connection at the inlet part to the analyzer and tried to fix it. The tube fell out and connected again quickly, but a short inhale of CO gas which was 38% concentration ensued. The researcher was hospitalized for 2 weeks to remove all CO trace in blood. After this accident, Tygon tube line sections for analyzers were all replaced with stainless steel tubes.

Hydrogen is actually a safe gas when proper instrumentation is applied, but it can cause a severe explosion if not properly dealt with. One explosion accident in Korea that ended with two deaths in 2019 happened by oxygen intrusion into the hydrogen tank [7]. More than 6% oxygen existed in the hydrogen buffer tank due to improper installation of the O_2 removing apparatus from H_2 gas as well as to higher O_2 existence inside the membrane system. Although hydrogen is safer than ordinary gases in terms of asphyxiation and toxicity, it is much more dangerous through explosion, which can bring greater damage. The combustion speed of hydrogen is fast enough to transfer from deflagration to detonation in which flame speed exceeds the sound velocity of 340 m/second.

These inherent features demand more investment in fabrication and in at least basic monitoring instruments. With these demands, it is not easy to implement gasification in a small-scale plant, considering the necessary instrumentation for preventing any unsafe operation.

4.4 Corrosion by sulfur-containing components in syngas

Syngas starting from natural organic feedstock should possess a certain concentration of sulfur-containing compounds, particularly H_2S in the gasification case. Biomass possesses a low level of sulfur and is not prone to corrosion as much as petroleum residues or coal. The most concerning problem occurs when H_2S is in liquid phase at the temperature below its boiling point in which H_2S will transform into sulfuric acid, and when the liquid phase stays inside the process. Sulfuric acid will corrode slowly and make pitholes to SUS 316 L stainless steel pipes, which are widely used under higher temperature conditions. Many gasification plant accidents from pipe failure happened at pipe sections where undesulfurized syngas stayed a long time (sometimes several months) by valve blockage at ordinarily unused sections.

Expensive pipe materials such as SUS 310, SUS 316, and frequent inert (typically nitrogen) gas purging process will definitely increase capital and operating costs.

4.5 Reduction of CO_2 emission

Most OECD countries promised to reach net zero CO_2 emission and carbon pollution-free targets by 2050, and other major CO_2 producing countries like China have also promised to attain the goal by 2060. Amine-based CO_2 capturing technologies are well proven in petrochemical industries, and recent developments include many upgraded sorbents that can lower the energy in recovering CO_2 after the capture step. The problem is that it typically suits large-scale plants and is costly when applied in small-scale gasification plants. Carbon credit prices were supposed to cover the involved cost of CO_2 removal, but reality has not progressed that way till now.

Figure 2 displays the carbon credit price in 2024 and the recommended carbon price in 2030 [8]. Most countries offer less than US$60/ton-$CO_2$, which is well below the value that can cover the required investment in CO_2 capture.

Technically it is already possible to reduce CO_2 emission up to 90% in large-scale operations. **Figure 3** shows the CO_2 capture level to meet the CO_2-emission regulation in US power plants [9]. In the figure, baseline (reference plant) means the conventional coal-fired steam power plant, and shift means the water-gas shift reaction. California state's CO_2 reduction requirement was 1100 lb-CO_2/MWh, and typical natural gas combined cycle (NGCC) power plants emit 770 lb-CO_2/MWh. When installing a one-stage shift reactor with an acid gas removal (AGR) facility to the existing IGCC power plant, CO_2 can be reduced 65%, and this is under the emission from NGCC, which means that the CO_2 emission level is below the natural gas case. Further installation of an additional shift reactor and advanced AGR, the CO_2 emission can drop to 10% of the current plant. These technologies were available already from the 2010s, but due to high cost, they were never applied to actual commercial plants. This emphasizes the necessity of an adequate carbon pricing strategy for CO_2 removal.

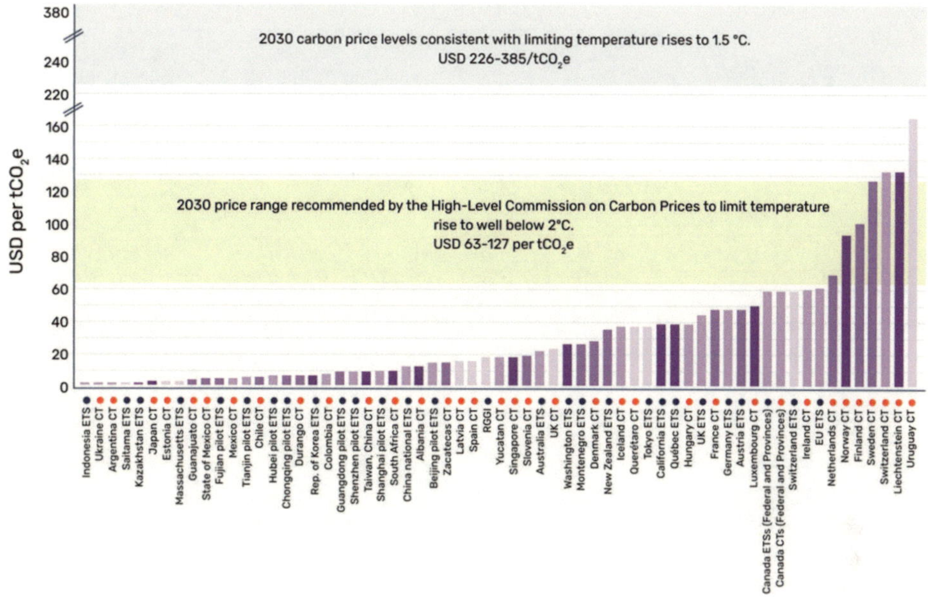

Figure 2.
World carbon credit price trend in 2024 and recommended 2030 carbon price levels [8].

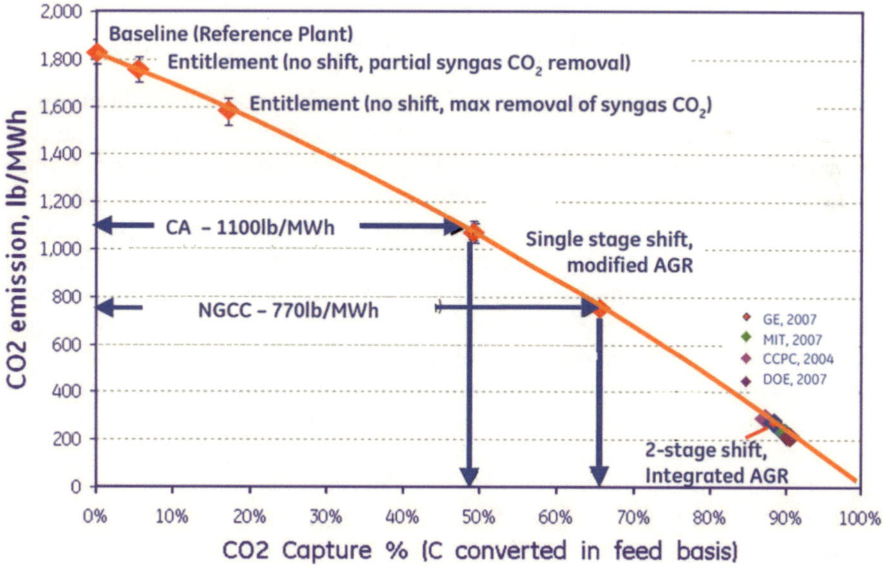

Figure 3.
CO_2 capture level to meet the CO_2-emission regulation in US power plants [9]. Note: CA (California), NGCC (natural gas combined cycle), AGR (acid gas removal).

What will be the cost in underground CO_2 storage, which is the most practical way to treat a large amount of CO_2? **Figure 4** demonstrates the costs involved in CO_2 capture and underground storage, which was estimated in 2007 [10]. The CO_2 capturing cost was estimated at that time as US$50/ton-$CO_2$, and in underground storage it was US$10/ton-$CO_2$. In total, it was US$60/ton-CO_2. Global technology development

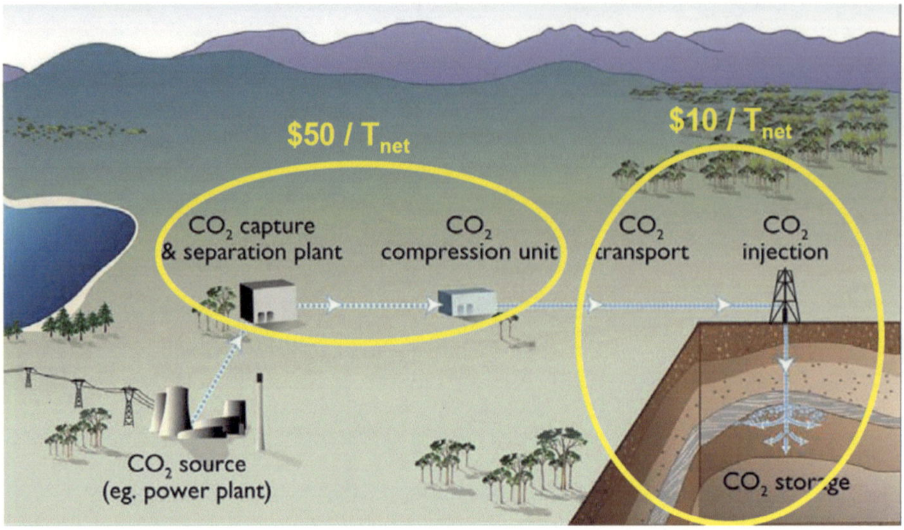

Figure 4.
Costs for CO_2 capture and underground storage, estimated in 2007 [10].

has targeted to reduce the cost to below US$20, but all efforts failed till now. In 2024 estimates, it will be even higher than US$60/ton-$CO_2$.

4.6 Competition with e-fuels and green hydrogen

In the coming decades that demand carbon-free and sustainable feeds, synthetic fuels and hydrogen from syngas have to compete with e-fuels such as e-methanol, e-SAF, and green hydrogen. Carbon-free transportation fuels are in great demands particularly in the passenger aviation market. In early 2025, a Swedish company announced ae 100,000 tons/year e-methanol plant, and a German company is said to be building a 110,000 tons/year green hydrogen plant using a proton exchange membrane (PEM) electrolyzer and electricity from wind power. A key issue is to produce lower greenhouse gas emission transportation (including aviation) fuels on a total lifecycle basis compared to the petroleum-based fuels.

If the process is chosen based on carbon intensity, gasification is not competitive because at least a few percent of CO_2 generation are inevitable in producing CO and H_2. Renewable electricity from wind and solar farms can be cheaper than fossil-fuel-fired power plants in many countries like in Spain, Australia, and countries where large desert areas are available. A gasification-based approach has some competitive hope when producing mass volume through large-scale plants and when applied to sustainable biomass/waste feeds.

4.7 Feasibility to produce clean hydrogen by gasification

Clean hydrogen is typically defined as hydrogen that emits CO_2 less than 4 kg-CO_2/kg-H_2 for production at the well-to-gate basis, and as the hydrogen that emits less than 2 kg-CO_2/kg-H_2 at the plant site.

As shown in **Table 4** [11, 12], a gasification plant using a fossil fuel like coal emits the highest CO_2 amount at almost 20 kg-CO_2/kg-H_2. Natural gas reforming, which is the most widely applied method to make H_2, produces almost 10 kg-CO_2/kg-H_2.

Process type for H_2 production	CO_2 emitting amount for H_2 production (kg-CO_2-eq/kg-H_2)	Reference
Fossil fuel gasification: coal gasification	19.85	[11]
Steam methane reformer (SMR)	10.85	[11]
Biomass gasification	5.22	[11]
Municipal solid waste gasification	4.99	[12]
Waste wood gasification	4.11	[12]
Water electrolysis using grid power	9.29	[11]
Photofermentation (biological method)	1.88	[11]

Table 4.
CO_2 emitting amount for H_2 production in various process types including gasification routes.

Biomass and waste gasification yields 4–5 kg-CO_2/kg-H_2. Generating H_2 by biomass/wastes gasification emits less CO_2 than water electrolysis using conventional grid electricity, which shows 9.29 kg-CO_2/kg-H_2. A recent trend in water electrolysis focuses on renewable electricity from wind and solar power to deal with CO_2 issues. Biological methods can produce H_2 at a much lower level compared to thermal technologies, but there are limitations in long processing time and in plant capacity.

Green hydrogen, which should be the ultimate target for future energy, is hydrogen produced through water electrolysis with 100% or near 100% renewable electricity, with close to zero greenhouse gas emissions, defined as less than 1 kg CO_2 per kg H_2 taken as an average over a 12-month period [13].

Gasification can reduce CO_2 emission by combining with additional carbon capture utilization and storage (CCUS) facilities. Whether the gasification process can compete with other renewable energy-based technologies remains to be answered. At least, hydrogen production through biomass and waste feeds indicates the direction to pursue. With technical advances on CO_2-capturing adsorbents in direct air capture (DAC) as well as in the post-combustion field, there are possibilities to overcome CO_2 barriers in the long run.

4.8 Delay in expected cost/scale of syngas-utilizing equipments

The key function of gasification is converting organic feeds into syngas in a relatively economical way. Syngas-utilizing equipment is so important that it can determine the fate of projects that involve gasification. From the early 2000s, it was hoped that cheap microgas turbines and fuel cells will find wide application as a distributed system. The microgas turbine was hailed as a future winner for the energy-distributed system where energy demand can be met by locally available energy sources. A future goal in microgas turbines is to manufacture it at a similar price to an automobile engine using gasoline or diesel, which was estimated as several hundreds of US dollars per small car engine in mass production. But this goal has not materialized yet, even after almost two decades. At this moment, internal combustion gas engines of several hundred kW scale are the most suited form of gasification plants in rural areas for distributed electricity, especially using biomass and wastes.

In the case of fuel cells, the expected price drop and bigger MW scale versions were not realized, even after decades of extensive global investments. Until now, it was actually false hope in syngas applications of MW scale by one unit. The integrated gasification fuel cell (IGFC) is still being pursued to exhibit above 60% of fuel-to-electricity efficiency. If fuel cells of MW scale come to market at competitive price, IGFC with carbon capture and storage (CCS) function can be a feasible option for the large-scale clean electricity power plants.

The only available option in using syngas to generate electricity has been through internal combustion gas engines. Gas engine has amassed application experience worldwide in the land filled gas (LFG) area up to MW scale, but its engine efficiency is not high and is still expensive due to limited global suppliers.

4.9 Difficulty in correcting mistakes after fabrication and installation

The same difficulty can appear in all cases of large-scale plants. But gasification, which works at high temperature and pressure like in IGCC, and with compact or complicated internal components, each step in fabrication should be verified carefully. In large mechanism industries, matured technologies have evolved more than 30 years of trial and error in fabrication and maintenance. The technology can be called matured with availability over 98% when most of the key mechanical components can be purchased in off-the-shelf and quick on-demand ways. Unfortunately, gasification plants did not reach that status, and there were many failure instances. Similar examples exist in the space shuttle *Challenger* disaster by a minor mistake initiated by an O-ring and the failure initiated by untightened washer nuts and loose screws in the James Webb Space telescope case.

Future application-pursuing people should pay attention to key items that had shown major failures in operation. One most common but disastrous wrong fabrication happened in refractory curing, refractory repair installation, and connecting the refractory-lined gasifier and syngas cooler. It happened in IGCC power plants in the United States and in Korea. It is a tremendously cost-intensive repair if a redo is decided on after installing the whole gasification block. Think about how to dismantle several hundred tons of intricately connected blocks and reinstall.

There are also many minor pieces of information on minimizing downtime in gasifiers. One example is that thermocouples for the hottest part in the gasifier should position not at the actual inside of the reactor, rather just a few centimeters away from the reactor inner wall, especially in refractory-installed gasifiers. Failure on this kind of minor item will cause long-term difficulties in monitoring the gasifier.

4.10 Difficulties in utilizing solid and liquid by-products

Inorganic components in feed became vitrified slag under high gasification temperature above the ash melting temperature, whereas they came out as ash under gasification temperatures below about 900°C as in fluidized-bed gasifiers. Because a fluidized bed should use heat-holding inorganic particles as a fluidization medium, ash particles after the gasification are better to use again inside the fluidized-bed gasifier.

Slags are supposed to be recycled as a benign construction material, but the reality is not that way. Even bricks made of gasification slags are not easy to sell in the open market. Bricks made from slags are utilized in public construction works such as in road blocks or in wall bricks along the river, at a much lower price.

Most people are reluctant to use the product, which originated from wastes in their own locale. Even though solid by-products from gasification passes all the regulatory requirements of heavy metal leaching test and of heavy metal contents, it is difficult to receive the actual market price. More novel methods of utilizing slags are in great demand.

Also for the liquid products, in the actual market, nobody is willing to pay the full price for pyrolysis oils as a replacement for petroleum-based products.

4.11 Economics

Gasification plants are relatively expensive compared to conventional plants that deal organic feedstock. This is mainly due to toxicity, explosiveness, and fire-prone tendency of syngas components, and the resultant instrumentation to cope with health and mechanical safety issues.

In the electricity generation area, an IGCC power plant is at least two times more expensive than the cost of conventional combustion-based steam power plants. **Table 5** illustrates recently constructed IGCC plants worldwide. The Kemper County plant case is exceptionally expensive since it tried to connect the world's first commercial-scale CCS facility into the IGCC scheme. Through the experience involved by the author in the Taean IGCC plant construction project, the construction cost of a O_2-blown entrained-bed type was estimated to go down as much as 20% in the next similar plant. Japan has constructed and is operating two ambitious air-blown entrained-bed IGCC power plants of 543 MW scale. Even with process complexities by air-blown gasification instead of O_2-blown, mainly due to economy of scale, they could reduce the construction cost to US$2700/kW.

Korea's actual case will be helpful to understand the construction cost in waste gasification. Construction cost of the government's official contract for a thermal pyrolysis plant using waste plastics feed that produces pyrolysis oils is US$600,000 per ton/day from 2025. This cost has more than doubled from the earlier unit construction cost guideline. The main background for the increase is for the cost in safety-related facilities. There have been many explosion and fire accidents in commercial pyrolysis plants that operated in Korea. Explosion accidents in kiln type pyrolyzers were the most common experience. The remaining pyrolysis gas has to be purged enough before igniting again during the operation, which is the same safety precaution that is applied in flare stacks. But accidents happen even with the obvious required precaution protocol. A typical waste gasification plant size of commercial scale is 50–150 tons/day. Considering the unit construction cost, this scale of plants requires about US$30 million to US$90 million, which is too high a number in ordinary civilian projects. In addition, operating cost in pyrolysis and gasification plants are about twice those of incineration plants of similar capacity. Incineration plants are typically at 300–1000 tons/day scale, which will guarantee cheaper operating cost simply due to economy of scale in addition to the maturity of technology.

IGCC plant	Reactor type	Capacity (MWe)	Construction cost (US$/kW)	Data year
Edwardsport IGCC, USA	O_2-blown entrained	618	4660	2014
Kemper County IGCC + CCS, USA	Circulating fluidized-bed	582	12,200	2017
Taean IGCC, Korea	O_2-blown entrained	300	3500	2018
Nakoso IGCC, Japan	Air-blown entrained	543	2700	2021
Hirono IGCC, Japan	Air-blown entrained	543	2700	2021

Table 5.
Comparison of construction cost per kW in currently working commercial IGCC plants.

In Korea's case, refuse plastic fuel (RFP) originated from waste plastics has been operated with reasonable profit through providing industrial steam by incineration. Competing by gasification method with this combustion-based steam-generating approach for the limited RPF source is not an easy task. When the final product is the same as the one from the combustion-based technology, it is difficult to compete in normal situations.

Note that the available waste amount for constructed gasification plants must be different, typically much reduced, whenever actual plant operation commences, by competition to get the good-quality wastes in the region. In the municipal wastes case, the most common value is 1 kg/person/day, which means that the MSW plant capacity for a 100,000 population city is 100 tons/day. In pyrolysis plants using waste plastics, key target plastics are polypropylene (PP) and polyethylene (PE), and they have to have less than 20% moisture and less than 1% chloride content. Locally available waste plastics of specific origin should be competitive in securing the sustainable amount continuously.

5. Choice of reactor type

The choice of gasifier depends on many factors in each region of the country. Note that the following information is basically the author's own opinion and that it cannot be correct in every situation. Gasifier types in the aspect of applying temperature were discussed in the earlier section on harnessing gasification benefits.

5.1 Fixed-bed

The Lurgi fixed-bed dry bottom gasifier has been successfully operated in South Africa's Sasol plants from the 1970s and in North Dakota's Synfuels plant from the 1980s. The fixed-bed type has a critical issue with produced tar from the recent aspect on environmental impact. The Lurgi technology has a maximum capacity of 1800 tons/day as the Mark V model.

The fixed-bed type can be suitable for gasification of biomass and MSW. Recent upgraded catalytic technology on tar cracking can solve the tar issue with proper investment. In MSW, Swiss-originated Thermoselect technology employs the compacting pretreatment in that wastes are compacted to remove any interior void space that can work as an insulator otherwise. Thermoselect technology for waste feeds has several commercial success records in Japan.

5.2 Fluidized-bed

If the bubbling type is employed, to guarantee good fluidization, uniform particle size and density in the feed are the most critical starting point. The transport type, in which fluidization gas velocity exceeds the terminal velocity of particles, has developed and demonstrated good performance in generating steam at the medium scale of application. It suits to adopt for biomass feed and also for MSW with proper pretreatment.

5.3 Entrained-bed

When the final preferable product is a gaseous type, entrained-bed in high temperature above the ash-melting temperature can serve better by minimizing liquid and solid by-products.

This type is most suitable for large scales of several thousand tons/day scale since it provides a very fast reaction time of a few seconds. The entrained-bed type suits for liquid feedstock of low grade such as petroleum residue oil and waste engine oil, because of feeding convenience and relatively uniform feed characteristics compared to other waste feeds.

But it is expensive and requires a heavy investment in instrumentation to cope with short response time to the fast reaction. With the technology matured and with more available off-the-shelf plant components with concurrent projects, this type can yield the economically competitive commercial example in IGCC-like big plants.

5.4 Plasma torch

Plasma can increase the inner side of the gasifier above 3000°C, and thus it is most suitable to treat dangerous waste feeds that can provide a very high tipping fee. The technology is still in the upgrading stage in many institutions and companies. Still, the high electricity amount required to fulfill the gasification is the most fundamental issue to obtain for wide applications.

The gasifier inner temperature distribution should be uniformly high enough to crack organics and melt inorganics if necessary. If any cold-spot area exists inside the slagging mode gasifier, it can lead to plugging in the slag-tap and causing lower performance. The volume that one plasma torch covers inside the gasification zone should have a certain limit; thus it has a considerable limitation in plant size scale-up.

The plasma type might be suitable for waste treatment of 30–150 tons/day scale, unfortunately not securing the benefit of economy of scale yet. The plasma type has great potential for medical wastes, by adopting the compact treatment system at the hospital site. Transporting harmful medical wastes can be minimized, and the only remaining residue will be slag for safe disposal.

6. Future direction

The most critical part in facilitating application of gasification must be cost effectiveness in commercial stages while reducing CO_2 intensity. Gasification technology needs to find rebirth as an efficient, ultra-clean method.

6.1 Feed uniformalization before gasification reaction

All chemical reactors are in a confined space, not in an open area. Fixed reaction space can be utilized ideally when uniform reactants enter the reactor and uniform products exit. When reactants are not uniform, reaction time for each component is different due to its reaction rate and the required residence time inside the reactor.

Gasification uses a reactor of confined volume due to its requirement of partial oxidation, which means no contact with ambient air. Feedstock of different elemental composition and of different configuration necessitates different reaction residence time inside the fixed volume of the reactor. When feed for gasification can be made to exhibit uniform performance inside the reactor, very high conversion is possible.

Gasification normally uses a low-value feedstock, which inherently contains a wide range of components. Wastes are the worst case as in municipal solid wastes. Actually, most waste treating facilities in major cities around the world use some sort of mixing and presorting tools to increase uniformity and to separate plastics and noncombustibles before thermal treatment. But more advanced techniques like torrefaction, pelletization, hydrothermal carbonization, and preliquefaction should be applied to improve feed uniformity. Torrefaction is a mild form of pyrolysis that removes volatiles and operates at temperatures between 200°C and 320°C. Hydrothermal carbonization treats high moisture content feeds like biomass with hot compressed water, which yields a coal-like product called hydro-char.

6.2 Minimizing required analytical tools

Gasification necessitates analytical tools in monitoring syngas composition and their trends along the process. Compared to a combustion-based system, which primarily needs to monitor combustion gases like CO_2, analysis tools on gasification syngas that contains harmful or explosive CO and H_2 are relatively expensive.

There are several points that can get information at some degree of accuracy without resorting to expensive analytical tools. Some examples are shown subsequently.

An entrained-bed gasifier typically operates at temperature higher than 1200°C. Under temperatures above 1250°C inside the reactor, a thermodynamic equilibrium calculation is sufficient to estimate in calculating the syngas composition from known composition of feed. From the author's experience, when the gasifier temperature gets higher than about 1450°C, the equilibrium thermodynamic calculation works pretty well. This means that outlet syngas composition can be easily estimated cheaply, even without accurate temperature data in a few degrees order. It is very helpful in preventing temperature overshoot that can damage the gasifier itself. This is also a very convenient, practical method in real plant operation. On-line syngas analytical apparatus is very expensive and requires heavy calibration and maintenance works. Companies manufacturing on-line analysis system are in many cases small companies, which has resulted in no more service on existing analytical systems after about 5–6 years.

A well-known method of estimating the temperature inside the hot gasifier is by measuring the CH_4 concentration. The CH_4 concentration is an effective indirect index of internal reaction temperature under the same operational schemes [14]. Methane concentration is sensitive to temperature variation as shown in **Figure 5**. When the gasification increases above 1400°C, CH_4 concentration drops sharply, and the relationship between CH_4 concentration and the gasifier temperature shows a relatively linear tendency. It can provide a good indicator for the internal temperature situation inside the gasifier, although it cannot guide the exact temperature number.

6.3 Zero-emission electricity generation and hydrogen production

In principle, zero emission is already possible technologically, if the involved cost is not considered. As discussed in an earlier section, gasification provides an environmentally benign way of converting N and S components in organic feedstock into sellable products, as shown in **Figure 6**. The zero-emission coal-utilizing IGCC concept in **Figure 6** was initiated in the 2000s with the lead taken by US the DOE through the FutureGen international consortium, but it did not reach actual implementation due to cost overrun. Without a firm guarantee on CO_2 revenue, it was destined to stop at a certain time. At that time, many regions around the world were eager to buy CO_2 for enhanced oil recovery (EOR) for pumping out crudes from underground, and it made sense to position this project near the EOR site and where cheap low-grade coal is available. This concept is still valid and can meet the requirement of carbon net-zero.

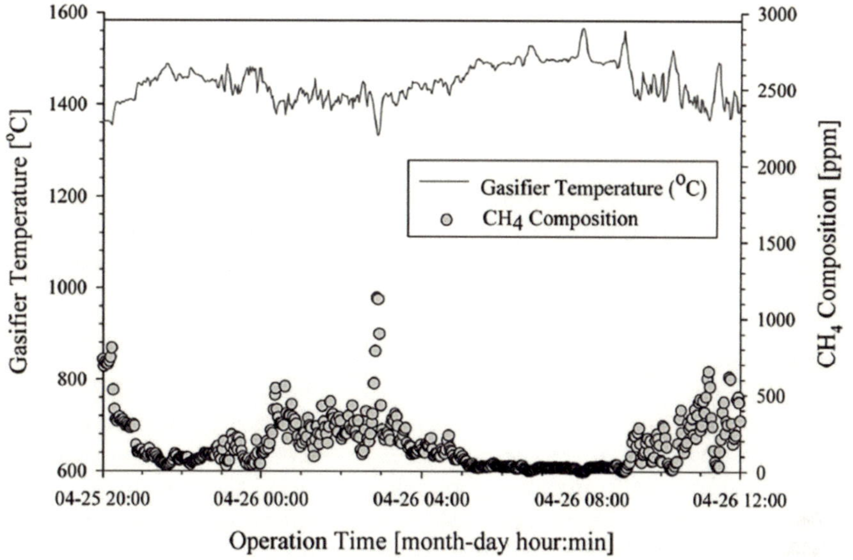

Figure 5.
Methane concentration with gasifier temperature in entrained-bed coal gasifier [14].

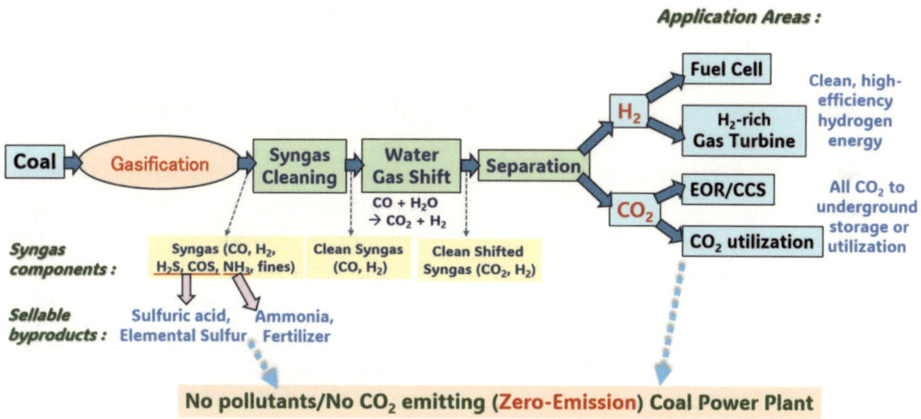

Figure 6.
Concept that developed in the 2000s on a zero-emission coal gasification plant for electricity and hydrogen.

After the water-gas shift reaction, feedstock will end up as gases containing only H_2 and CO_2, with minor other components. Hydrogen will be used as a feed to fuel cell and to gas turbine to produce electricity. CO_2 will be used as a medium instead of conventional steam(water) in EOR for pumping out crudes from underground and will be permanently (more than several hundred years) at the underground EOR site, replacing underground void volume that exiting crudes leave when extracted.

A large amount of hydrogen is required to change the fuel in steelmaking to hydrogen. In this case, a zero-emission coal or petroleum residue fired gasification plant that equips CCS function might be the most feasible option to supply hydrogen.

6.4 Modular plant for localized distributed gas-based society

The modular fabrication approach has been applied in other industries like housing and oil and gas offshore construction for quite a long time. It is a natural direction for reducing manufacturing cost, through standardization of instrumentation and operation protocol. In other competing sectors of e-fuels, e-hydrogen, and biogas utilizing plants, modular type is already widely employed. In the plant size of 50–150 tons/day, this approach is a must-have item that the gasification field has to garner for competing in unit price and as a way for standardizing remote operation that can reduce operating cost.

6.5 Application of big-data statistical tools

One of the most effective and practical methods in process monitoring involves comparing the process temperature/pressure/syngas composition results with the simulated data obtained from thermodynamic calculation and from big-data accumulation. Temperature and pressure measuring devices are cheapest among signal sensors. Actually, temperature and pressure are two most basic ways to describe the state of a system in thermodynamics. It is the identical concept as applied to thermodynamics, which is based on four measurable properties of temperature, pressure, volume, and mass.

The cheapest items in process monitoring are pressure and temperature sensors. The best way in getting the gasification system cheaper while monitoring the system fully might be maximizing the usage of these kinds of cheap tools. In principle, if process variables can be estimated by these pressure and temperature values through mathematical calculations and/or big operational data accumulation, it will reduce the number of expensive measuring apparatus required.

Appropriate operational big data at the certain process of fixed feedstock and machine learning models can assist in gasification plant operation with much fewer personnel and less instrumentation that will yield more economical competitiveness. Gasification plants cannot be operated by novice operators because of the inherently toxic and dangerous syngas composition. Experienced operators are much needed but are getting more difficult to hire in rural areas. Reducing direct operation by highly skilled operators with operation-supporting software should be the direction that is continuously pursued. Remote-sensing software is already widely used in plant engineering. Even though it requires more refined upgrading to prevent malicious outside intrusion, merging with big-data software can lead to more effective gasification operation.

7. Conclusions

Perceptions on gasification coexist in two ways. On one hand, gasification matured from the 1930s in Germany and from the 1950s in South Africa, particularly in coal gasification and conversion-to-chemicals areas. On the other hand, it is regarded as far from a matured state in dealing with environmental and CO_2 reduction areas. Under the competitive circumstances with products made by renewable electricity, which does not emit CO_2, existing gasification technologies are not in a good position to persuade markets. They also have to compete with biogas that is based on the biological route, which is slow but can be much cheaper at small scales. Gasification itself is a basic tool to make syngas from organic feedstock, and the final application depends on the associated apparatus and conversion routes. The gasification technologies need to revamp and incorporate available apparatus and software that have evolved to cheaper and high-efficiency performance.

In this chapter, benefits of employing gasification and application areas were summarized. Furthermore, inherent and fundamental issues around gasification were discussed with feasible future directions on gasification.

With recent market requirements in using biomass and wastes, gasification technologies can evolve to meet the energy and chemicals requirements which specify carbon neutrality and environmentally benign operation.

Acknowledgements

This research was supported by a grant (2050000079) from the "Development of Demonstration-Scale Hydrogen Production Technology Using Pet-coke" program funded by the Ministry of Land, Infrastructure, and Transport of the Korean government.

Author details

Yongseung Yun
Plant Engineering Division, Institute for Advanced Engineering, Yongin-si, Gyeonggi-do, Republic of Korea

*Address all correspondence to: ysyun@iae.re.kr

IntechOpen

© 2025 The Author(s). Licensee IntechOpen. This chapter is distributed under the terms of the Creative Commons Attribution License (http://creativecommons.org/licenses/by/4.0), which permits unrestricted use, distribution, and reproduction in any medium, provided the original work is properly cited.

References

[1] Yun Y. The 2nd Korea IGCC and Coal Gasification Symposium. Suwon, Korea: Korea Gasification Association; 2018

[2] Xia C et al. Prospect of near-zero-emission IGCC power plants to decarbonize coal-fired power generation in China: Implications from the GreenGen project. Journal of Cleaner Production. 2020;**271**:122615-122634

[3] Kamall R. Technology Status Report–Coal Liquefaction. US Department of Energy; 1999. Available from: https://www.netl.doe.gov/sites/default/files/netl-file/technology-status-coal-liquefaction.pdf

[4] Ko H. KOWEPO Taean IGCC Operation Updates. Gasification User's Association. Colorado Springs, Colorado, USA: Electric Power Research Institute; 2018

[5] Eni to use Nextchem's gasification technology to produce circular methanol and hydrogen. Hydrocarbon Processing. 2025. Available from: https://www.hydrocarbonprocessing.com/news/2025/02/eni-to-use-nextchems-gasification-technology-to-produce-circular-methanol-and-hydrogen/

[6] Parraga J. Polygeneration with biomass-integrated gasification combined cycle process: Review and prospective. Renewable and Sustainable Energy Reviews. 2018;**92**:219-234

[7] Gangneung Science Complex Hydrogen Tank Explosion Incident. Namuwiki website. 2025. Available from: https://en.namu.wiki/w/%EA%B0%95%EB%A6%89%20%EA%B3%BC%ED%95%99%EB%8B%A8%EC%A7%80%20%EC%88%98%EC%86%8C%ED%83%B1%ED%81%AC%20%ED%8F%AD%EB%B0%9C%20%EC%82%AC%EA%B1%B4

[8] The World Bank Open Knowledge Repository. State and Trends of Carbon Pricing 2024. 2024. Available from: https://openknowledge.worldbank.org/entities/publication/b0d66765-299c-4fb8-921f-61f6bb979087

[9] Shilling N. IGCC: Its Role in Solving the Carbon Puzzle, Gasification Technologies Conference 2007. San Francisco, USA: Gasification Technology Council; 2007

[10] Kaldi J. CCS Activities in Victoria. CO_2 CRC Workshop. Australia: CO_2 CRC; 2007

[11] Yagmur Goren A et al. A comprehensive review on environmental and economic impacts of hydrogen production from traditional and cleaner resources. Journal of Environmental Chemical Engineering. 2023;**11**:111187

[12] Lui J et al. Life cycle assessment of waste-to-hydrogen systems for fuel cell electric buses in Glasgow, Scotland. Bioresource Technology. 2022;**359**:127464

[13] Green Hydrogen Organization (GH2). The Green Hydrogen Standard. 2022. Available from: https://gh2.org/ourinitiatives/gh2-green-hydrogen-standard

[14] Yun Y, Chung SW. Gasification of an Indonesian subbituminous coal in a pilot-sacle coal gasification system. Korean Journal of Chemical Engineering. 2007;**24**(4):628-632

Section 2

Hydrogen Production by Gasification

Chapter 2

Blue Hydrogen Production in Pilot and Demo-Scale Plants through Pet-Coke Slurry Gasification, WGS, and PSA Units

Seung Jong Lee, Yongseung Yun, Jin Wook Lee and Seok Woo Chung

Abstract

Syngas from gasification is converted into hydrogen and CO_2 through the water-gas shift (WGS) reaction, and the subsequent pressure swing adsorption (PSA) separation produces high-purity hydrogen while capturing high-concentration of CO_2. The collected CO_2 can be utilized, at least as a feedstock for dry ice, which is in short supply in many countries. There are still many under-utilized low-grade feedstocks, such as pet-coke, biomass, and waste. These feedstocks can serve as good starting materials for producing economically feasible hydrogen for fuel cell power generation and hydrogen-fueled automobiles. Gasification technology is effective at breaking down low-grade feedstocks into syngas ($CO + H_2$). However, in pet-coke gasification, there is an additional challenge due to its high sulfur content, which necessitates extensive desulfurization. Specifically, to use hydrogen in fuel cells, the residual sulfur concentration in hydrogen must be reduced to below 0.1 ppm. Over the last 3 years, a 2-ton/day pilot plant has been operated to produce high-purity hydrogen from pet-coke, using gasification, desulfurization, WGS, and PSA units. Based on the results from the pilot plant, a scaled-up 20-ton/day demonstration plant was constructed, yielding preliminary results toward the ultimate goal of producing blue hydrogen. Operation results show that sulfur containing impurities can be reduced to less than 0.1 ppm, and hydrogen of 99.9% purity can be obtained after WGS and PSA units. Several business models were proposed to utilize the developing process by module approach.

Keywords: blue hydrogen, pet-coke, gasification, syngas, WGS, PSA

1. Introduction

As a result of global warming, the issue of decarbonization has accelerated in recent years, and countries around the world are continuing their efforts

IntechOpen

toward carbon neutrality by expanding the share of renewable energy, increasing hydrogen usage, and utilizing various sustainable methods [1–3]. With regard to renewable energy, each country has different inherent resources, and for nations like South Korea, where renewable energy resources are limited, there are various constraints and limitations in resource utilization. Therefore, in addition to renewable energy, it is essential to optimize the use of other carbon-free energy sources such as hydrogen, carbon capture utilization and storage (CCUS), and nuclear power to promote the achievement of carbon neutrality [4, 5]. As part of a specific strategy to accelerate global carbon neutrality, the South Korean government proposed the CFE (Carbon Free Energy) Initiative in 2023 [6, 7]. During the 28th Conference of the Parties to the United Nations Framework Convention on Climate Change (COP28), the proposal for the global expansion of the "CFE Initiative" was introduced, and major countries such as the United Kingdom, France, the Netherlands, Saudi Arabia, the UAE, and Japan have officially expressed their support and interest [8, 9]. CFE refers to energy that produces no carbon emissions throughout its entire life cycle, from production to storage, transportation, and utilization. Renewable energy and nuclear power, both of which produce no carbon emissions, can be used as carbon-free energy sources. However, for hydrogen to qualify as a carbon-free energy source, no greenhouse gas emissions should occur throughout its entire life cycle. In line with this principle, green hydrogen, blue hydrogen, and turquoise hydrogen are classified as types of comprehensive clean hydrogen.

Recent climate-conscious initiatives are encouraging the production of at least blue hydrogen instead of gray hydrogen [10–12]. In blue hydrogen production, CO_2 is captured and stored underground through Carbon Capture & Storage (CCS) or utilized as sellable products via Carbon Capture and Utilization (CCU) [4]. Gray hydrogen remains the dominant technology supplying most of the hydrogen currently in use. Currently, global production of green hydrogen is limited, so in the short term, the use of blue hydrogen and turquoise hydrogen is expected to rise [13]. However, in the medium to long term, the use of green hydrogen is expected to gradually increase [14].

As a practical solution for large-scale clean hydrogen production, hydrogen can be produced through gasification or reforming of fossil fuels such as natural gas or coal, while the CO_2 generated during the production process is removed and utilized through CCUS. The societal demand for CO_2-free H_2 is pushing the transition of existing gray hydrogen technologies, at least to blue hydrogen, with the ultimate goal of advancing toward green hydrogen (**Figure 1**) [15].

Utilizing low-grade resources such as municipal solid waste, waste plastic, and pet-coke can provide the foundation for large-scale hydrogen production, making a technical review of this approach essential [16–18]. Pet-coke, a byproduct of crude oil refining, currently has a high carbon content and thus high calorific value, while it contains a high sulfur content [19–21]. When sulfur content is elevated, it is primarily used as boiler fuel or road paving material. Producing hydrogen from pet-coke first requires converting it into gaseous form under the oxygen-limited gasification condition, generating the syngas primarily composed of carbon monoxide and hydrogen. This process demands the design, fabrication, and operational technologies for a plant that includes advanced purification, reforming, conversion, and high-purity hydrogen separation. This chapter presents the results of constructing and operating pilot-scale and demo-scale facilities for blue hydrogen production using pet-coke as a feed (**Figure 2**).

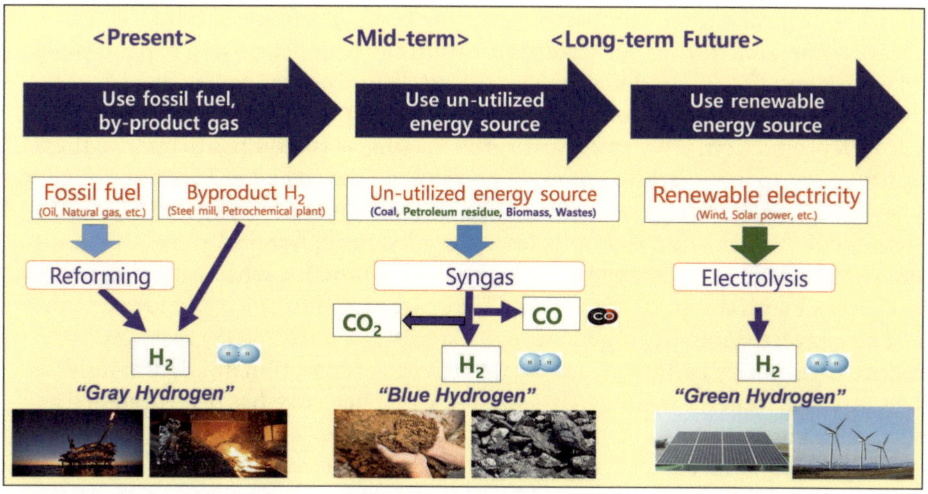

Figure 1.
Prospects for the production and application of clean hydrogen.

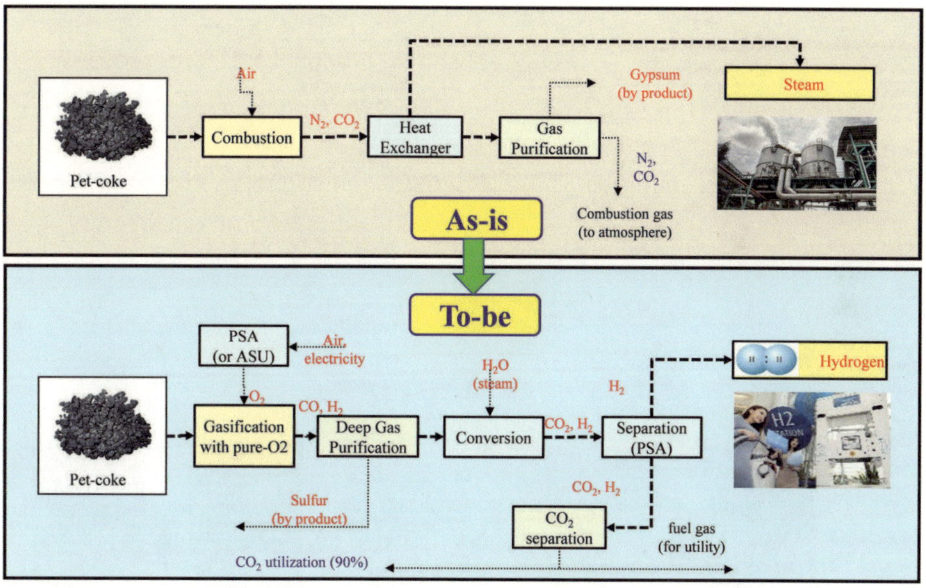

Figure 2.
Comparison of as-is and to-be process schemes for pet-coke utilization.

2. Configurations of hydrogen production system using pet-coke

Pet-coke contains about 85–92% fixed carbon and 5–8% sulfur. When gasified, the organic materials (C, H, O, N, S, etc.) in pet-coke are converted, with the C, H, and O components transforming into syngas primarily composed of carbon monoxide and hydrogen, while the N and S components are converted into harmful gases such as NH_3, H_2S, and COS. Due to its high calorific value and with the purpose of more hydrogen content in syngas, gasification of pet-coke can be more effective when using a wet slurry-type feeding method, where pet-coke is mixed with water to form a slurry before gasification, compared to the dry-feeding type gasification. The harmful gases generated during pet-coke gasification can cause corrosion in plant pipes and equipment [22–24] and may poison the catalysts used for converting syngas into hydrogen [25–29]. Therefore, to ensure efficient process operation and stable hydrogen production, advanced purification of syngas is required to remove these harmful gases to levels below 1 ppm. In particular, due to the 1 ~ 7 wt% sulfur content in pet-coke, gasification generates 0.3 ~ 5 vol% sulfur containing gases, H_2S and COS. Advanced desulfurization technologies are needed to economically and efficiently remove these high-concentration acidic gases. Additionally, when gasifying pet-coke slurry mixed with water, approximately 50% carbon monoxide is produced, which must be converted into hydrogen through a water-gas shift (WGS) reaction. To produce high-purity hydrogen, a separate hydrogen separation unit is also necessary. Therefore, a system for hydrogen production using pet-coke has been designed as shown in **Figures 3** and **4**.

The gasification feedstock, petroleum coke, was sourced from USA. Liquid oxygen with a purity of over 99.5%, supplied by Air Products, was vaporized and used as the oxidant for the gasification process. To protect the analytical instruments, purge

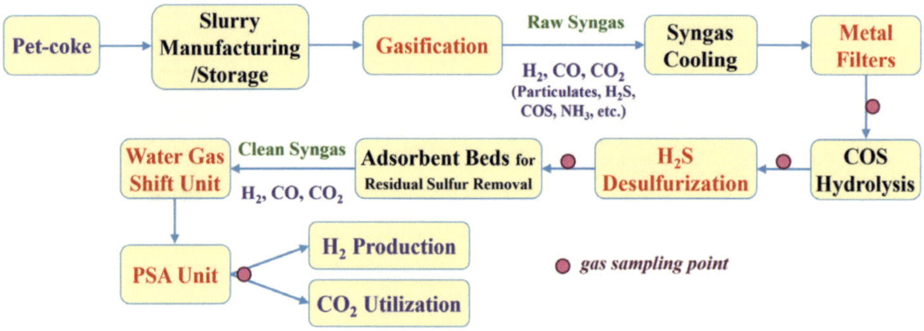

Figure 3.
Process flow diagram of blue hydrogen production system using pet-coke.

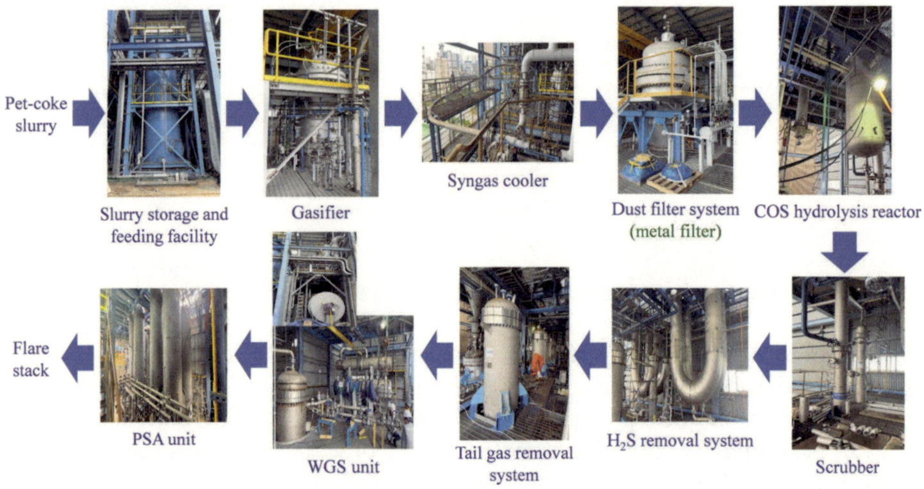

Figure 4.
Process flow for blue hydrogen production in demo-scale plant using pet-coke.

nitrogen was supplied by Air Products as liquid nitrogen with a purity of 99.999% and vaporized prior to use.

Prior to each gasification experiment, the petroleum coke was analyzed by SGS Korea, a KOLAS-accredited (Korea Laboratory Accreditation Scheme) testing laboratory. The analytical results were verified and confirmed before proceeding with the gasification operations.

For the demo-scale plant, a slurry was produced using 20 ton/day of pet-coke, which was then gasified to produce syngas at a rate exceeding 2000 Nm^3/h. The harmful gases in the syngas, such as H_2S, were reduced to below 0.1 ppm, and HCl/ NH_3 were reduced to below 0.2 ppm before being sent to the WGS system. In the WGS system, the water-gas shift reaction converted over 99% of the carbon monoxide in the syngas into hydrogen. The pressure swing adsorption (PSA) system was designed to produce 3 ton/day of hydrogen with a purity greater than 99%. The detailed configuration of the system for blue hydrogen production using pet-coke is shown in **Figure 5**.

Pilot plant typically operates to verify the product characteristics under different operating conditions and for various feedstock in each key process components. In this study, operation duration in the pilot plant was normally 8 ~ 12 hours after reaching stable feeding condition. After securing enough information from the pilot plant on gasification, WGS performance, and PSA system, then demo plant of typical ten times scale was envisioned. Typically demo plant operates to accumulate the operating data for all involved facilities and optimize any technical issues, which act as design basis for the future commercial-scale plant, as well as to provide more accurate economic results before commercial project investment. Demo facility in this study has started to operate from 2023. Normal operation time in demo plant is minimum 1 day under steady-state operating condition.

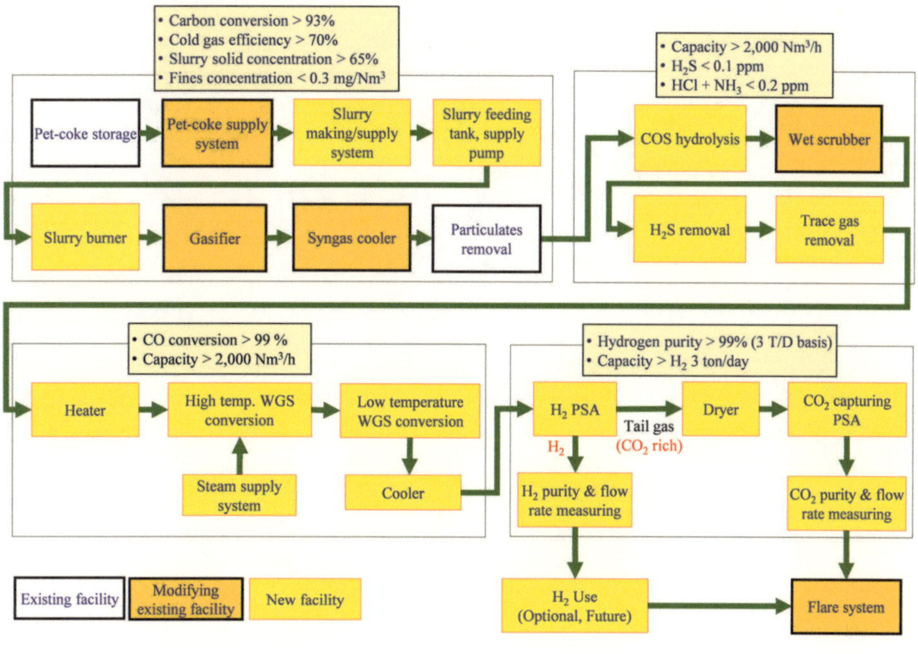

Figure 5.
Detail block diagram of demo-scale blue hydrogen production system using pet-coke.

To analyze the syngas produced from gasification, a sampling system was configured as shown in **Figure 6**, and the gas analysis locations are indicated in **Figure 3**. At each analysis location, the water trap, particulate filter, and high-pressure regulator indicated in **Figure 6** were individually installed on-site. Gas analysis was performed by switching between the analysis locations as needed.

Real-time analysis of syngas and H_2 was conducted using two Nova Prime Syngas analyzers (MRU, Germany), while H_2S and COS concentrations were monitored using an online GC system (Thermo Fisher Scientific Inc., USA). The measured data were continuously displayed and recorded through the monitoring screen in the control room, as shown in **Figure 7**.

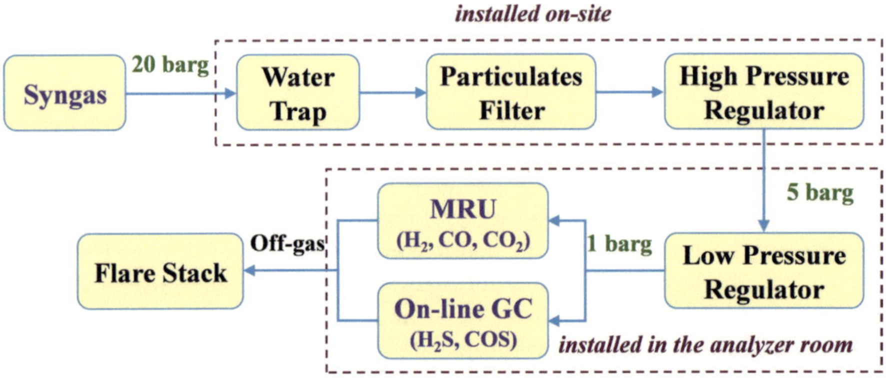

Figure 6.
Block diagram of gas pretreatment system for analysis.

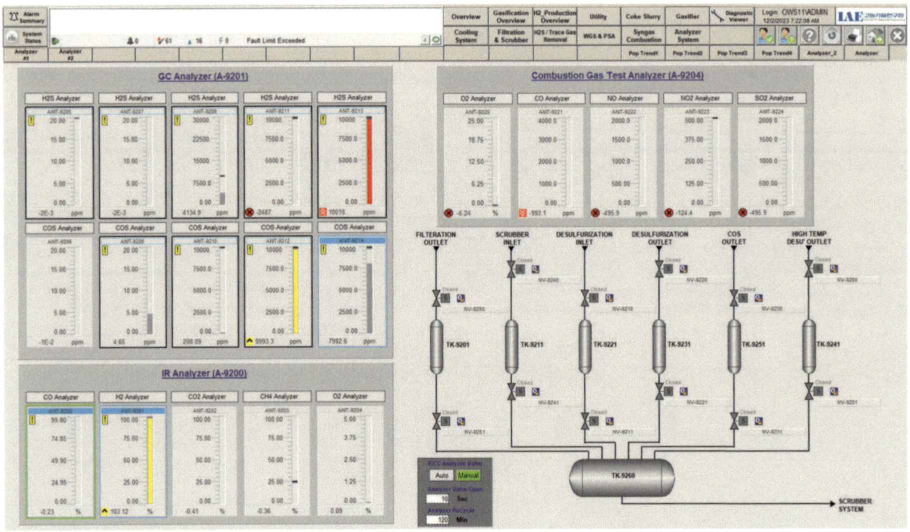

Figure 7.
Monitoring screen for gas analysis system.

3. Pilot-scale hydrogen production system using pet-coke

3.1 Configuration of pilot-scale hydrogen production system

The pilot-scale facility has been utilized to analyze the gasification characteristics of coal or pet-coke at a rate of 2 ton/day and to enhance gasification performance by evaluating factors such as carbon conversion rate, cold gas efficiency, and others. Additionally, to produce hydrogen from pet-coke, the facility equipped to generate hydrogen from 10 ~ 30 Nm^3/h of syngas, a portion of the approximately 200 Nm^3/h of syngas produced at the pilot scale. The block diagram of the pilot-scale blue hydrogen production system is shown in **Figure 8**, and the overview of actual pilot-scale facility is provided in **Figure 9**.

The 10 ~ 30 Nm^3/h hydrogen production system using slip-stream of syngas, shown in **Figure 10**, consists of high-purity acidic gas purification equipment, CO conversion (WGS) equipment, and hydrogen separation (PSA) equipment. Since the pilot-scale facility is a small-scale experimental setup and does not include steam production equipment like commercial plants, a separate steam generator has been installed to provide the necessary steam for the water-gas shift (WGS) reaction in the WGS system.

3.2 Operation results of pilot-scale hydrogen production system

Hydrogen production using pet-coke was carried out in a pilot-scale facility. As shown in **Table 1**, a slurry with a viscosity of 150 ~ 300 cP was produced using approximately 2 ton/day of pet-coke and then gasified. Feeding rate of pet-coke in slurry form was about 3 ton/day. The gasifier operated at the temperature of 1450 ~ 1500°C range and at the pressure of 15 bar. The oxygen required for the gasification of pet-coke was supplied as high-purity oxygen vaporized from liquid oxygen, at a weight ratio of 0.75 ~ 0.85 relative to the pet-coke slurry weight.

As a result of gasifying pet-coke under the conditions mentioned above, syngas was produced with CO content of 46 ~ 48%, H_2 content of 25 ~ 26%, and CO_2

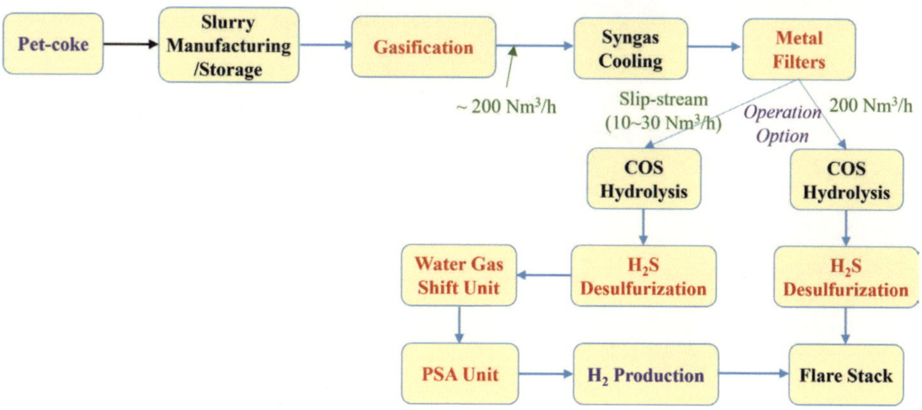

Figure 8.
Block diagram of pilot-scale hydrogen production system.

Figure 9.
Overview of pilot-scale plant (2 ton/day).

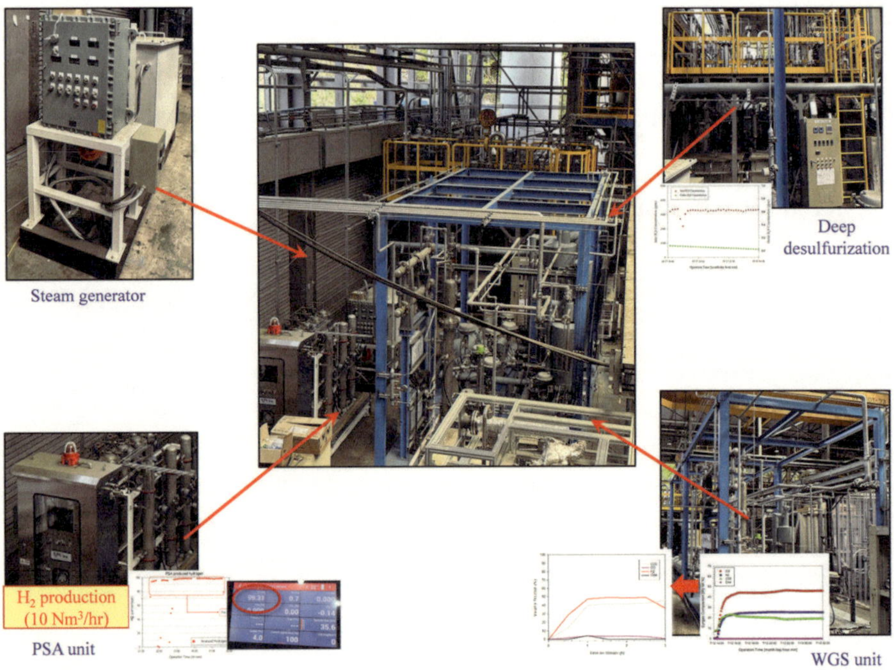

Figure 10.
H_2 production system using slip-stream syngas (10 ~ 30 Nm^3/h) from pilot-scale plant.

Item	Operating condition/Data
Slurry feeding amount	120 ~ 140 kg/h (dry pet-coke ~2 ton/day)
Pet-coke concentration in slurry feed	60 ~ 64 wt%
Pet-coke feed proximate analysis	Fixed carbon 87.4 wt%, VM 9.6 wt%, Ash 0.17 wt%, moisture 0.9 wt%, Sulfur 2.7 wt%, gross HV 8337 kcal/kg
Slurry feed viscosity	150 ~ 300 cP
Gasifier operation temperature	1450 ~ 1500°C (maximum temp. area)
Gasifier operation pressure	15 bar
Oxidant	Pure oxygen from liquid oxygen
O_2/Slurry weight ratio	0.75 ~ 0.85

Table 1.
Characteristics of 2 ton/day (in slurry 3 ton/day) pilot-scale gasification plant.

content of 16 ~ 18%, as shown in **Figure 11**. The cold gas efficiency was found to be between 64 and 67%, and the carbon conversion rate was confirmed to be between 93 and 95.6%.

In the case of COS, 200 ~ 250 ppm was produced, most of which was hydrolyzed into H_2S in the COS hydrolysis system, reducing its concentration to below 0.5 ppm. For H_2S, approximately 3500 ~ 4400 ppm was produced and subsequently reduced to below 0.2 ppm in the acidic gas removal system, as shown in **Figure 12**.

In the WGS system, as shown in **Figure 13**, the syngas temperature at both the inlet and outlet of the high-temperature reactor (HTS) was maintained between 200 and 500°C, while the syngas temperature at both inlet and outlet of low-temperature reactor (LTS) was maintained between 120 and 180°C. As shown in **Figure 14**, hydrogen with a purity of 99% was produced after passing through the PSA system.

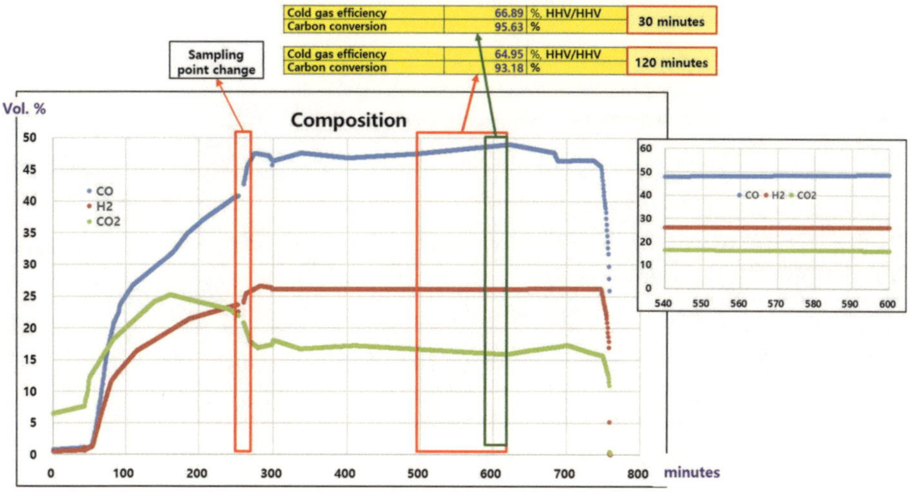

Figure 11.
Typical gasifier operation result of pilot-scale plant.

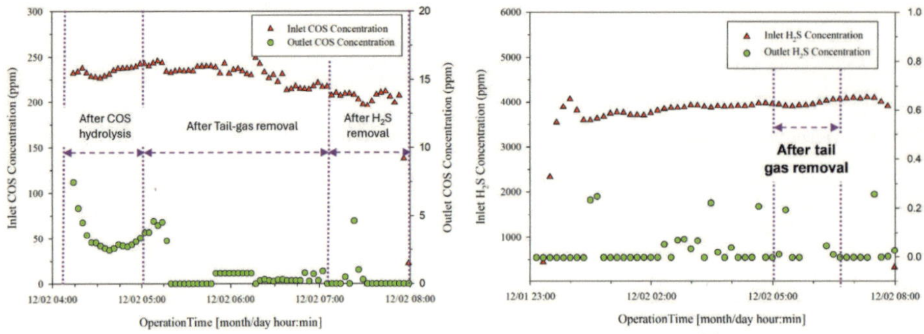

(a) (b)

Figure 12.
Operation results of acid gas removal process in pilot-scale plant. (a) COS concentration at inlet and outlet of hydrolysis (b) H_2S concentration at inlet and outlet of H_2S absorber.

(a) (b)

Figure 13.
Operation results of WGS process in pilot-scale plant (a) Inlet and outlet temperature of HTS reactor (b) Inlet and outlet temperature of LTS reactor.

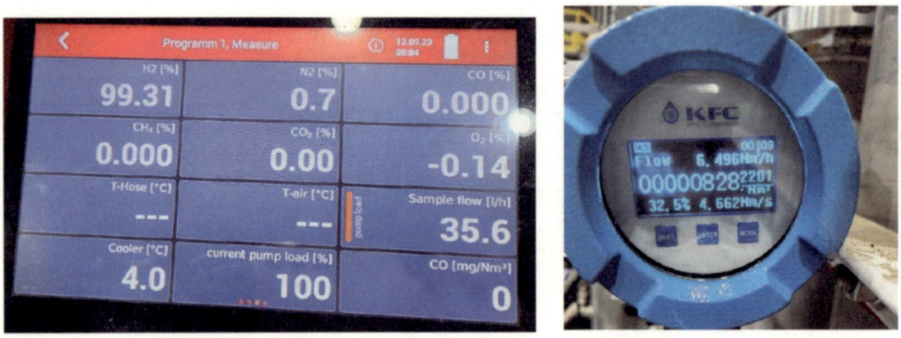

(a) (b)

Figure 14.
Operation results of PSA process in pilot-scale plant (a) Pressure trends of H_2 in PSA tower (b) H_2 gas flow rate measured.

4. Demo-scale hydrogen production system using pet-coke

4.1 Configuration of demo-scale hydrogen production system

The gasification facility installed at the demo-scale plant is a non/partial slagging entrained-bed gasifier, originally designed for the gasification of sub bituminous coal which contains a high volatile component. Since 2015, it has been operating under the conditions of maximum 1500°C temperature and maximum 20 bar pressure. The gasifier had partially modified for pet-coke gasification and has been in operation for this purpose since 2023. An overview of the demo-scale plant is shown in **Figure 15**.

The demo-scale plant has equipped with a gasifier located on the 2nd floor up to the 4th floor. The 4th floor houses also the particulates removal system, the 3rd floor contains the purification system, the 2nd floor is dedicated to the WGS system, and the 1st floor accommodates the PSA system. An overview of each floor and the control room is shown in **Figure 16**.

The particulates removal system utilizes metal filters to capture and remove fly ash after the gasification. Due to long-term operation since 2015, some of the dust collection filters, supports, and backwash modules became corroded, and new components were fabricated and installed.

The H_2S removal system uses an iron chelate solvent to directly decompose and eliminate H_2S through a chemical reaction. The system composed of an absorption tower, regeneration tank, supply tank, and solvent supply pump. The absorption tower operated using a scrubbing method, and a demister was installed at the tower's rear to prevent iron chelate solvent from carrying over to the downstream section.

Figure 15.
Overview of demonstration scale plant for hydrogen production using pet-coke.

Figure 16.
View of each floor and control room in demo-scale plant.

The WGS system consists of a high-temperature shift reactor (HTS), a low-temperature shift reactor (LTS), a heat exchanger, and separate steam generator that designed to supply necessary steam for WGS reaction. The PSA system composed of four adsorption–desorption units, along with instruments such as solenoid valves and pressure gauges.

4.2 Operation results of demo-scale hydrogen production system

After the construction of syngas purification, WGS, and PSA systems, the demo-scale plant was operated for hydrogen production using pet-coke slurry as a feed. The gasifier was designed to process up to 32 ton/day of pet-coke slurry, operating at temperatures ranging from 1430 to 1500°C and a pressure of 20 bar. Oxygen required for pet-coke gasification was supplied as high-purity oxygen, vaporized from liquid oxygen, at a weight ratio of 0.65 ~ 0.75 relative to the pet-coke slurry weight. Note that this required oxygen to slurry ratio is lower than the one used in pilot-scale plant which was 0.75 ~ 0.85. In general, required oxygen diminished a certain value in larger scale gasification reactors due to lower heat loss in the larger diameter vessel (**Table 2**).

During the operation of the demo-scale plant, the syngas and hydrogen produced are not stored separately but are instead combusted in the ground flare stack, where the flames are contained inside the stack. Combustion of syngas and hydrogen in the flare stack is continuously monitored to ensure stable combustion and meeting the environmental regulations before being emitted. The operational control screens for the gasifier, dust collection system, desulfurization system, WGS system, and PSA system during operation are shown in **Figure 17**.

Item	Operating condition
Slurry feeding amount	1330 ~ 1350 kg/h (pet-coke slurry ~32 ton/day)
Gasifier operation temperature	1430 ~ 1500°C (maximum temp. area)
Gasifier operation pressure	20 bar
Oxidant	Pure oxygen from liquid oxygen
O_2/Slurry weight ratio	0.65 ~ 0.75

Table 2.
Characteristics of 20 ton/day (in slurry 30 ton/day) demo-scale gasification plant.

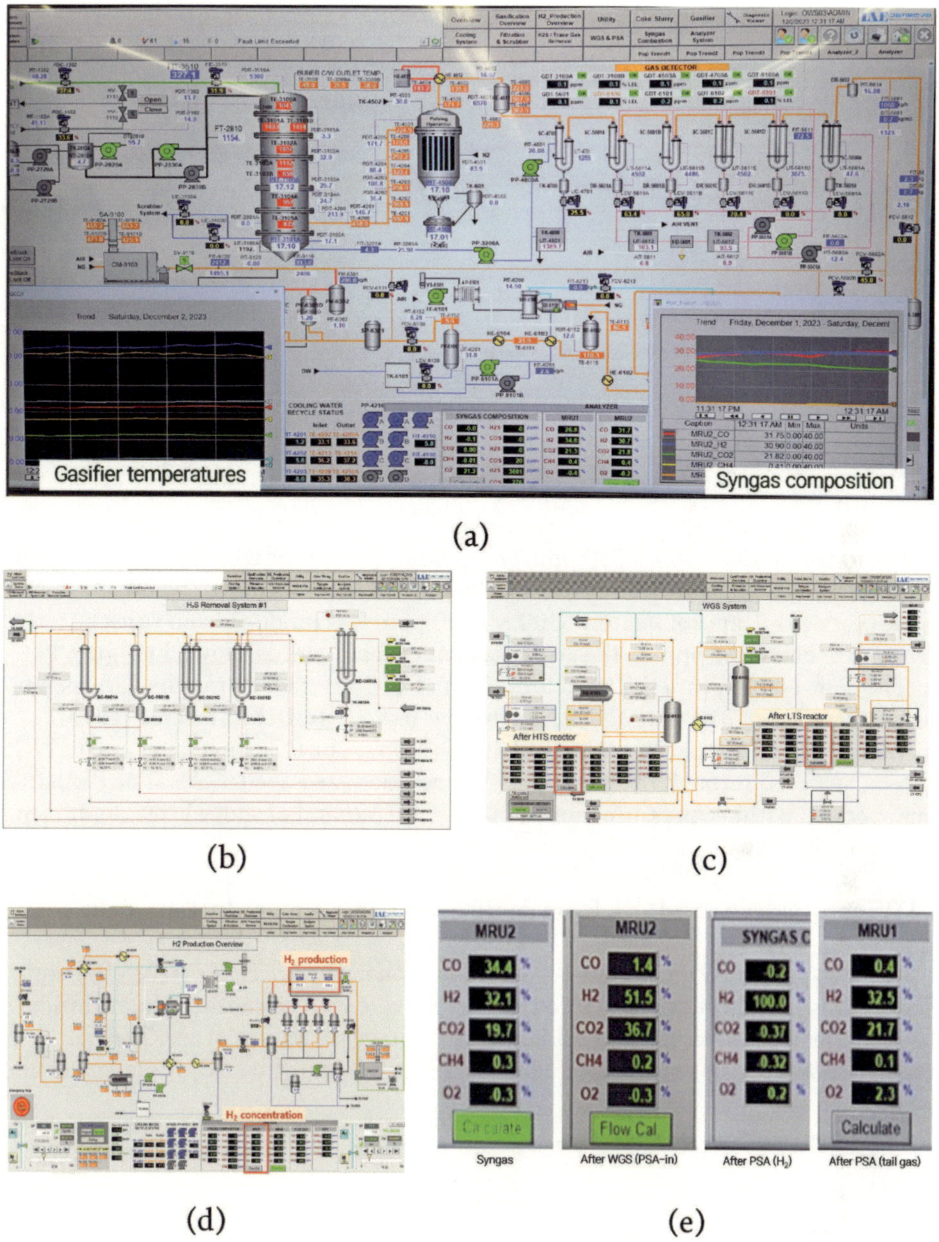

Figure 17.
Control screen views for key facilities and gas compositions monitored in control room of demo-scale plant (a) Overview control screen for gasifier-syngas cooling-particulate removal-desulfurization systems (b) Desulfurization using Iron chelate (c) WGS for CO conversion (d) PSA for H_2 separation (e) Gas compositions through process.

Result of gasifying pet-coke in the demo-scale plant exhibited the syngas composition of 40 ~ 45% CO, 23 ~ 25% H_2, and 15 ~ 20% CO_2. Note that CO_2 concentration in syngas was 19.7 and 36.7% after WGS system, as shown in **Figure 18**. This CO_2 can be more processes in CO_2-PSA to yield more than 99% CO_2 purity which will be sufficient for transport and for further industrial usage.

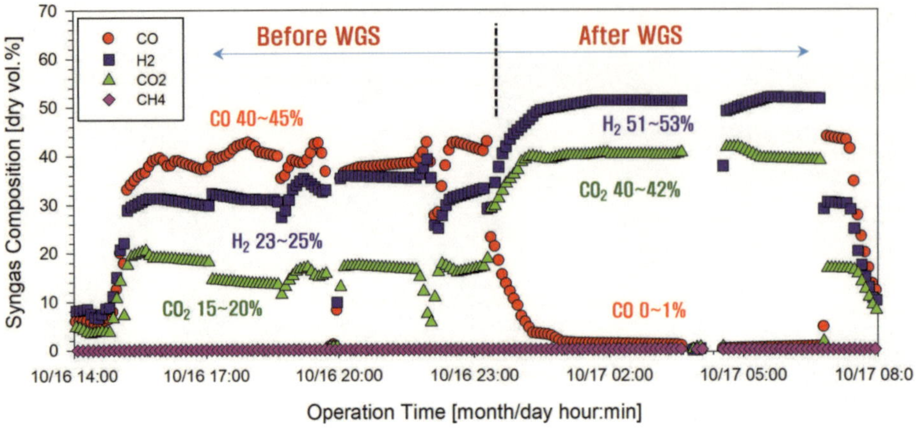

Figure 18.
Operation result of syngas composition in demo plant. (20 ton/day).

In the case of COS, 360 ~ 560 ppm was produced, most of which was hydrolyzed into H_2S in the COS hydrolysis system, reducing its concentration to below 0.5 ppm. In the case of H_2S, approximately 5300 ~ 6200 ppm was produced and subsequently removed to below 0.1 ppm in the acidic gas removal system, as shown in **Figure 19**.

The high-temperature shift reactor (HTS) of WGS system maintained an operating temperature of 360 ~ 460°C, while the low-temperature shift reactor (LTS) operated within a temperature range of 150 ~ 300°C. The CO conversion rate in WGS system, achieved through the water-gas shift reaction, averaged 97.3%. The PSA system operated within a pressure range of 0.01 ~ 9.0 bar and produced final hydrogen at the purity exceeding 99% (**Figures 20** and **21**).

4.3 Ways to use captured CO_2 from demo-scale hydrogen production plant

The CO_2 generated during hydrogen production using pet-coke can be efficiently converted into high-purity CO_2 through the PSA system. The captured high-purity CO_2 can then be transported for underground storage or industrial use. Typical required CO_2 purity is at least 95% which is the case for pipeline CO_2 transport, while

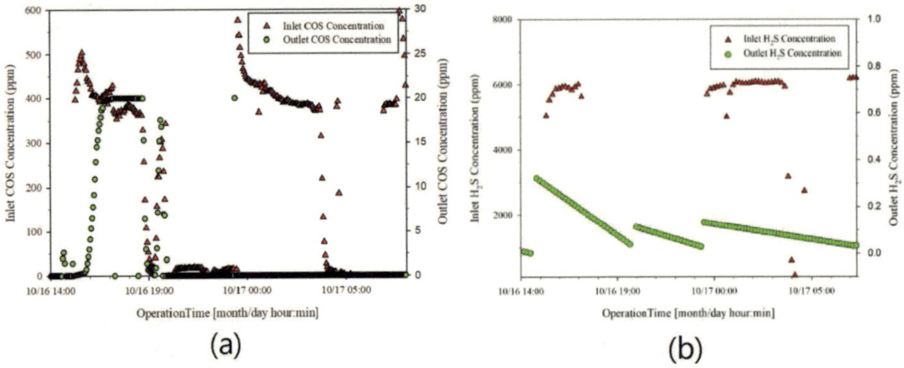

Figure 19.
Operation results of acid gas removal process in demo-scale (20 ton/day) plant (a) COS concentration at inlet and outlet of hydrolysis reactor (b) H_2S concentration at inlet and outlet of H_2S absorber.

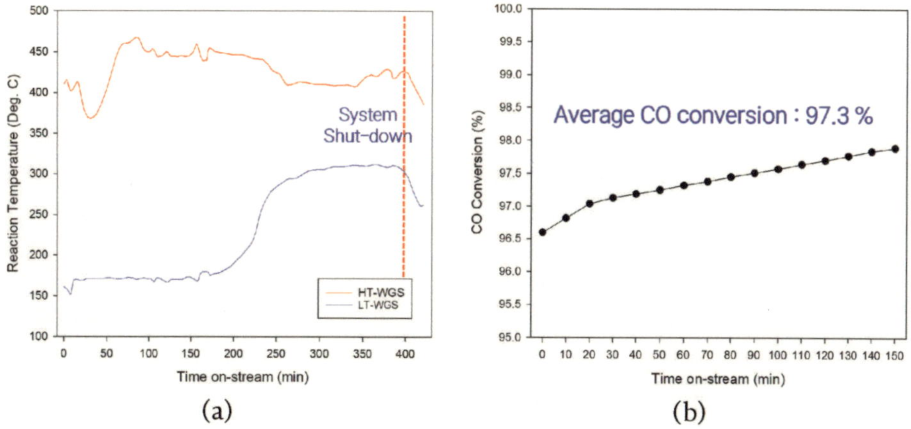

Figure 20.
Operation results of WGS process in demo-scale (20 ton/day) plant (a) Temperature of HTS and LTS reactor (b) CO conversion ratio.

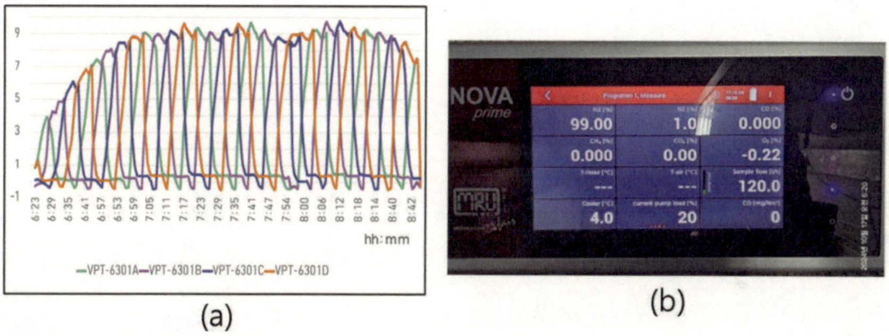

Figure 21.
Pressure swing trends and final H_2 purity at PSA system in demo-scale (20 ton/day) plant (a) Pressure trends of H2 PSA tower (b) H2 99.0%.

liquefied CO_2 transport by tanker trucks or ships should meet the CO_2 purity requirement of higher than 99.5% [11]. CO_2 concentration after WGS system ranges around 35 ~ 40%, and another PSA unit will work for concentrating CO_2 up to higher than 99% level.

In South Korea, there are no suitable sites for CO_2 storage, meaning that a significant amount of captured CO_2 must be processed industrially. However, due to the challenges in scaling up chemical conversion methods for large volumes of CO_2, commercial applications of this approach are still not feasible. Therefore, CO_2 mineralization may present the most viable alternative for the industrial recycling of CO_2. Although the composition of petroleum coke may vary, a 20-ton/day demo-scale hydrogen production plant is expected to produce approximately 56 tons of CO_2, which will be captured and utilized through mineralization.

CO_2 mineralization involves using materials containing CaO, such as recycled aggregates, recovered water from ready-mix concrete, or industrial by-products like power plant ash or refinery ash, to mineralize captured CO_2. The mineralized CO_2 can then be used in the production of low-grade construction materials, such as bricks, road paving materials, and insulators, as shown in **Figure 22**. The pilot-scale CO_2 mineralization plant developed by Daewoo Engineering & Construction Co. is shown in **Figure 23**.

Figure 22.
Concept of CO_2 mineralization to utilize captured CO_2.

Figure 23.
Pilot plant for $CaCO_3$ construction material production (Daewoo Engineering and Construction Co., CO_2 treating capacity 1 ton/day).

4.4 Business models using demo-scale hydrogen production plant

When using one hydrogen fuel cell car, which is available from Korean Hyundai and Japanese Honda, Toyota, assuming 10,000 km mileage a year, about 100 kg/year of hydrogen (99.99% purity) is required. Current hydrogen fuel cell car can run

100 km with 1 kg of hydrogen. The demo plant in this study of 32 ton/day of pet-coke slurry capacity can produce about 2 tons/day of 99.99% purity hydrogen which in turn can support 1000 hydrogen cars, assuming one car runs 200 km/day. At this time of most countries, current demo plant scale in this study might be sufficient until the full hydrogen society arrives.

Demo-scale hydrogen production plant using pet-coke was designed and constructed to produce and process syngas at a scale of 2500 Nm^3/h. It includes modules for advanced desulfurization, water-gas shift, and hydrogen separation (PSA), which can be directly applied for the commercialization of various syngas plants using wastes, biomass syngas/biogas plants, and natural gas reforming plants. Additionally, as shown in **Figure 24**, the demo-scale plant is designed to utilize both syngas produced from pet-coke and syngas produced at the nearby Taean commercial IGCC power plant. The syngas produced from pet-coke will be generated for 2 months annually, while 2500 Nm^3/h, which accounts for 1.3% of the syngas produced at the Taean IGCC power plant, will be supplied for 10 months. If the syngas from pet-coke proves to be more economically advantageous than that produced at the Taean IGCC power plant, the number of operational days for pet-coke syngas will be extended beyond 2 months.

Core equipments being developed have a gas processing capacity of 2500 Nm^3/h, featuring identical advanced desulfurization, water-gas shift, and hydrogen separation (PSA) modules. As shown in **Figure 25**, this scale is suitable for direct application in medium-sized syngas plants using wastes or biomass and in natural gas reforming sectors. The 2500 Nm^3/h gas volume corresponds to the capacity of a typical 50 ton/day waste gasification plant (equivalent to the municipal waste generated by a city with a population of 500,000), a 50 ton/day combustible biomass gasification plant, a 600 Nm^3/h natural gas reforming plant, and a 10 MW renewable electricity-based P2G hydrogen plant. Once standardized modules are fully developed, they can be commercialized across various sectors. In particular, the gasification of waste plastics for hydrogen production or medical waste gasification plants can be scaled to match the required gas capacity, and the potential applications for each module are vast.

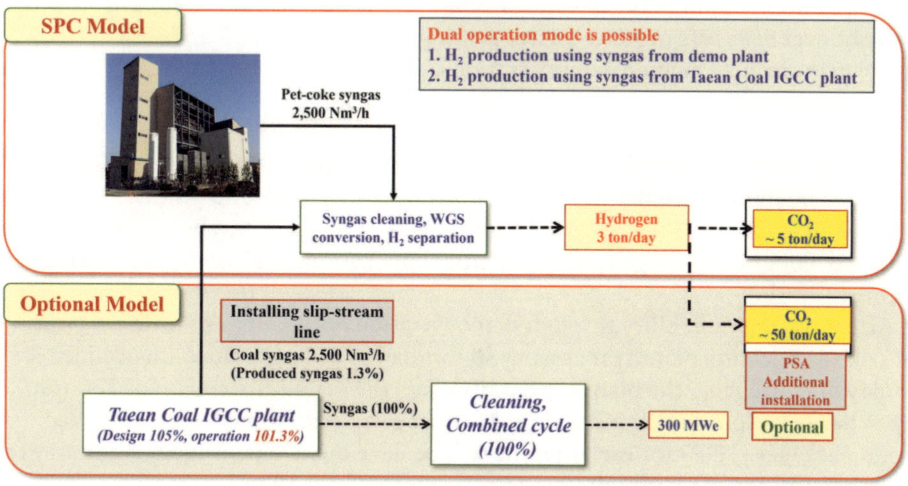

Figure 24.
Business models using demo-scale plant and plant modules for hydrogen production.

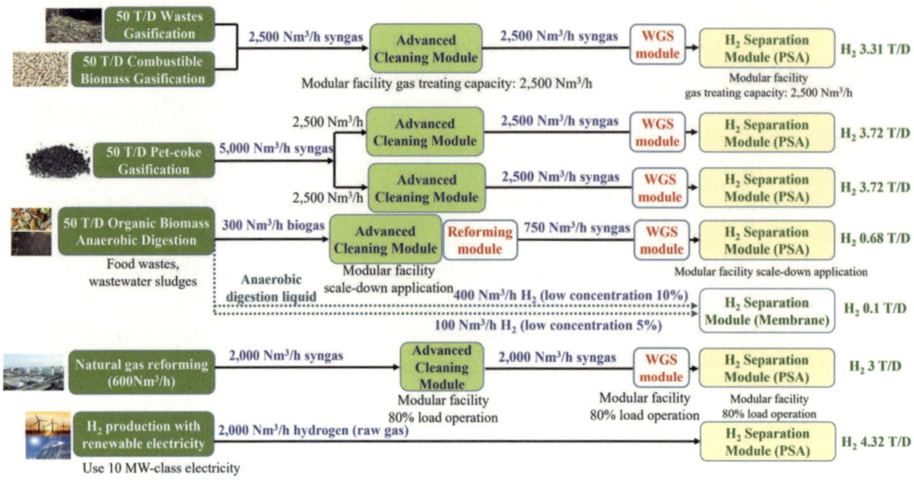

Figure 25.
Module approach for acid gas removal, WGS, and PSA sections of demo-scale plant.

5. Conclusions

To develop the hydrogen production plant using pet-coke, a pilot-scale plant was first constructed and operated. In the advanced syngas purification system, high concentration of acidic gas (H_2S) exceeding 3000 ppm, which were produced during the pet-coke gasification, was removed to below 0.1 ppm. After passing through the WGS and PSA systems, hydrogen with a purity exceeding 99% was produced.

After confirming the production of high-purity hydrogen in the pilot-scale plant, the process was scaled up to a demo-scale plant, where 20 ton/day of pet-coke were supplied to produce 3 ton/day of hydrogen. The demo-scale plant produced syngas with the following composition: CO 40 ~ 45%, H_2 23 ~ 25%, CO_2 15 ~ 20%, H_2S 5300 ~ 6200 ppm, and COS 360 ~ 560 ppm. Similar to the pilot-scale plant, high-concentration acidic gases (H_2S) above 3000 ppm were removed in the demo-scale plant to below 0.1 ppm in the advanced syngas purification system. Hydrogen with a purity of over 99% was produced after passing through the WGS and PSA systems. Future plans in 2025 include demonstration operations to ensure the reliability of the demo-scale plant.

Hydrogen produced from pet-coke is targeted to achieve a purity of 99.9% for use as feedstock for fuel cell and transportation fuels applications. Depending on the final demand requirements, PSA facilities will be upgraded to produce hydrogen with purities of 99.99% or 99.999%. Additionally, high-purity carbon dioxide will be produced alongside hydrogen, with potential applications such as dry ice production or mineralization.

After ensuring reliability through demonstration operations at the demo-scale pet-coke gasification plant (processing 30 ton/day of pet-coke slurry to produce 3 ton/day of hydrogen), the plan is to detail design the scaled-up version of 200 ton/day of pet-coke (producing 30 ton/day of hydrogen). Additionally, standardized design packages (PDP) for each module will be developed, enabling market entry for pet-coke gasification plants with capacities ranging from 50 to 300 ton/day. This will allow for further scale-up and market expansion into hydrogen production plants utilizing biogas or waste-derived syngas.

Acknowledgements

This research was supported by a grant (2050000768) from the "Development of Demonstration-scale Hydrogen Production Technology using Pet-coke" program funded by the Ministry of Land, Infrastructure, and Transport of the Korean government.

Appendix

1. **Figures A1** and **A2** show the process flow diagram of the pilot-scale gasification facility, the control room view, and the utility supply and gas analysis systems that required for plant operation.

2. As depicted in **Figure A3**, the control room not only manages the operation of the facility but also monitors the status of the equipment through CCTV. Additionally, thermal monitoring infrared cameras were installed around the gasifier to continuously check for any leaks or heat losses during the operation.

3. The H_2S removal system installed in the demo-scale plant is shown in **Figure A4**.

4. The WGS and PSA systems installed at the demo-scale plant are shown in **Figures A5** and **A6**, respectively.

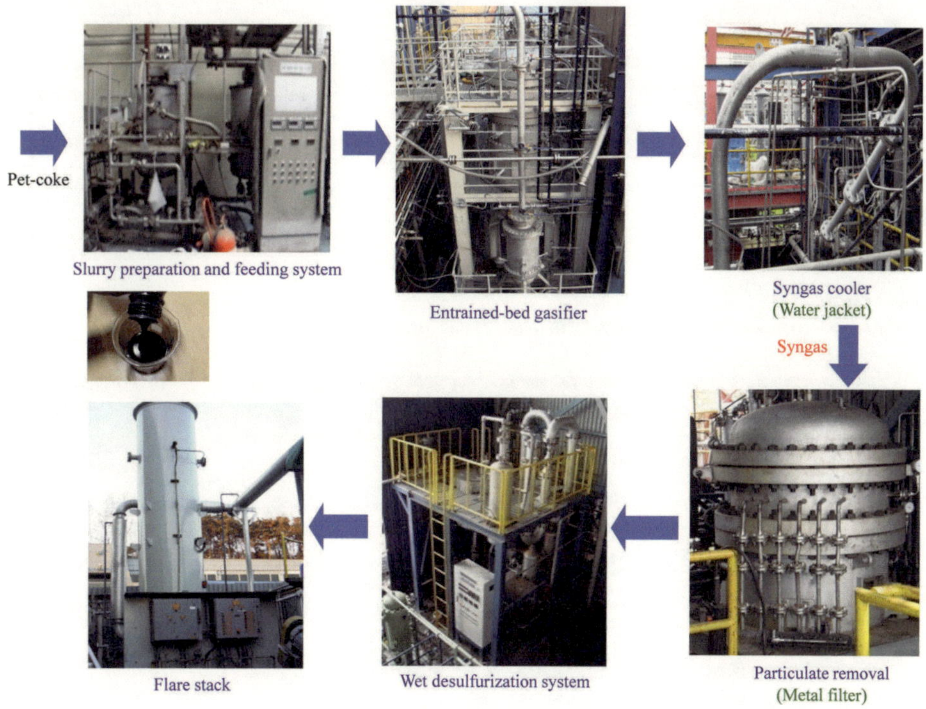

Figure A1.
Process flow of pilot-scale plant (2 ton/day).

Figure A2.
Utilities supply, syngas analysis systems, and control room view of pilot-scale (2 ton/day) plant.

Figure A3.
View of control room and monitoring system in demo-scale plant.

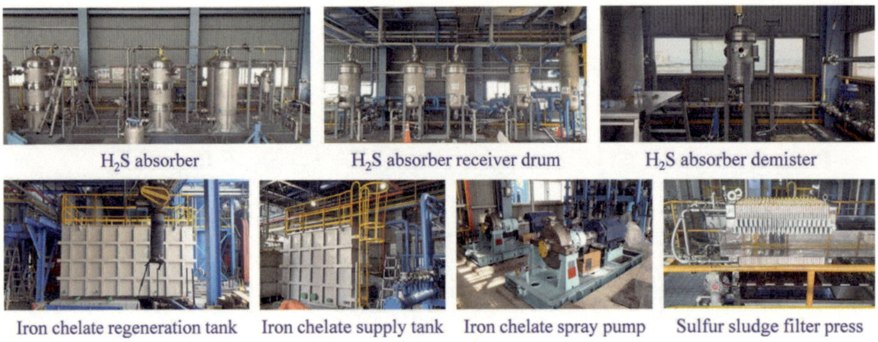

Figure A4.
Key components of H_2S removal system (syngas treating capacity 1600 ~ 2500 Nm^3/h) in demo-scale plant.

Figure A5.
Key components of WGS system (syngas treating capacity 1600 ~ 2500 Nm^3/h) in demo-scale plant.

Figure A6.
Fabrication and installation of PSA system (syngas treating capacity 1600 ~ 2500 Nm^3/h) in demo-scale plant.

The appearance of the flames in the flare stack and the monitoring of this process via CCTV in the control room are shown in **Figure A7**.

The 3D design view for the $CaCO_3$ material production demonstration plant is provided in **Figure A8**.

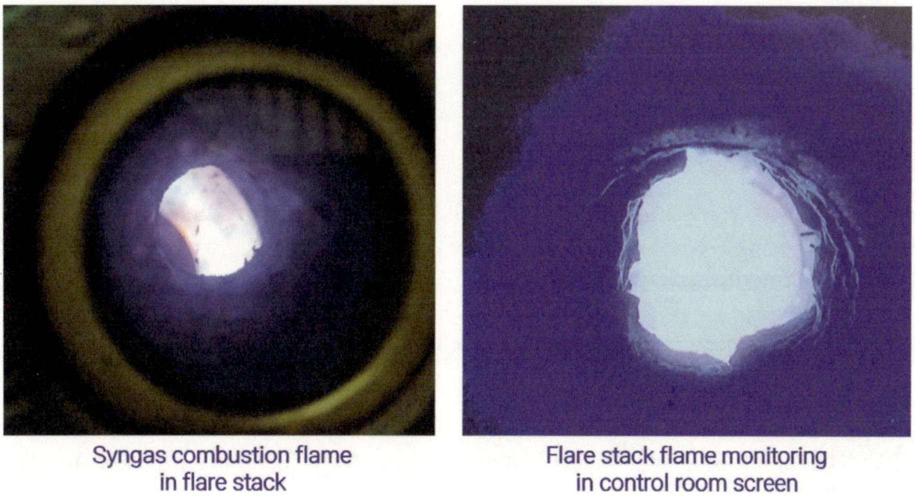

Figure A7.
Syngas combustion view inside flare stack in demo-scale plant.

Figure A8.
Proposed test plant 3D design for $CaCO_3$ construction material production using captured CO_2 (IAE, Ulsan, Korea, CO_2 treating capacity 5 ton/day).

Author details

Seung Jong Lee*, Yongseung Yun*, Jin Wook Lee and Seok Woo Chung
Plant Engineering Division, Institute for Advanced Engineering, Baegam-Myeon, Cheoin-Gu, Yongin-Si, Republic of Korea

*Address all correspondence to: sjlee@iae.re.kr and ysyun@iae.re.kr

IntechOpen

© 2025 The Author(s). Licensee IntechOpen. This chapter is distributed under the terms of the Creative Commons Attribution License (http://creativecommons.org/licenses/by/4.0), which permits unrestricted use, distribution, and reproduction in any medium, provided the original work is properly cited.

References

[1] IEA. Net Zero by 2050: A Roadmap for the Global Energy Sector. 2021. Available from: https://iea.blob.core.windows.net/assets/deebef5d-0c34-4539-9d0c-10b13d840027/NetZeroby2050-ARoadmapfortheGlobalEnergySector_CORR.pdf

[2] Bhuiyan MMH, Siddique Z. Hydrogen as an alternative fuel: A comprehensive review of challenges and opportunities in production, storage, and transportation. International Journal of Hydrogen Energy. 2025;**102**:1026-1044. Available from: https://www.sciencedirect.com/science/article/pii/S0360319925000382

[3] Zjou Y. Worldwide carbon neutrality transition? Energy Reviews. 2023;**2**:100026. Available from: https://www.sciencedirect.com/science/article/pii/S2772970223000135

[4] Hydrogen Council. Hydrogen Insights 2023. Hydrogen Council & McKinsey & Company; 2023. Available from: https://hydrogencouncil.com/en/hydrogen-insights-2023/

[5] Shukla PR, Skea J, Slade R, Khourdajie AA, Hasija A, Malley J, et al. Climate Change 2022: Mitigation of Climate Change. IPCC; 2022. Available from: https://www.ipcc.ch/report/ar6/wg3/

[6] Kim S. Trends and Future Challenges of the Carbon-Free Energy (CFE) Initiative Energy Focus; 2024. pp. 52-62. Available from: https://www.keei.re.kr/keei/download/focus/ef2403/ef2403_50.pdf

[7] Ministry of Foreign Affairs of the Republic of Korea homepage. Clean Energy. Available from: https://www.mofa.go.kr/eng/wpge/m_5657/contents.do

[8] Office for Government Policy Coordination. Launches the Carbon-Free (CF) Coalition, Leading Discussions on International Norms. 2023. Press Release. Available from: https://www.2050cnc.go.kr/base/board/read?boardManagementNo=3&boardNo=2483&page=&searchCategory=&searchType=&searchWord=&menuLevel=3&menuNo=1397

[9] Ministry of Trade, Industry and Energy. Maximize the Use of all Carbon Free Energy for Carbon Neutrality. 2024. Press Release. Available from: https://www.2050cnc.go.kr/base/board/read?boardManagementNo=43&boardNo=2878&searchCategory=&page=52&searchType=&searchWord=&menuLevel=2&menuNo=92

[10] IEA. Global Hydrogen Review 2024. 2024. Available from: https://www.iea.org/reports/global-hydrogen-review-2024

[11] DNV. Energy Transition Outlook 2024. 2024. Available from: https://www.dnv.com/energy-transition-outlook/download/

[12] Ochu ER, Braverman S, Smith G, Friedmann J. Hydrogen Fact Sheet: Production of Low-Carbon Hydrogen. Columbia University Energy Policy Institute; 2021. Available from: https://www.energypolicy.columbia.edu/sites/default/files/pictures/HydrogenProduction_CGEP_FactSheet_052621.pdf

[13] Boretti A, Pollet BG. Hydrogen economy: Paving the path to a sustainable, low-carbon future. International Journal of Hydrogen Energy. 2024;**93**:307-319

[14] Ishaq H, Dincer I, Crawford C. A review on hydrogen production and

utilization: Challenges and opportunities. International Journal of Hydrogen Energy. 2022;**47**:26238-26264

[15] Ahn SY, Kim KJ, Kim BJ, Hong GR, Jang WJ, Bae JW, et al. From gray to blue hydrogen: Trends and forecasts of catalysts and sorbents for unit process. Renewable and Sustainable Energy Reviews. 2023;**186**:113635

[16] Rauch R, Kiros Y, Engvall K, Kantarelis E, Brito P, Nobre C, et al. Hydrogen from waste gasification. Hydrogen. 2024;**5**:70-101. DOI: 10.3390/hydrogen5010006

[17] Rubinsin NJ, Karim NA, Timmiati SN, Lim KL, Isahak WNRW, Pudukudy M. An overview of the enhanced biomass gasification for hydrogen production. International Journal of Hydrogen Energy. 2024;**49**:1139-1164

[18] Rezende TTG, Venturini OJ, Palacio JCE, Oliveira DC, Santos DJS, Lora EES, et al. Technical and economic potential for hydrogen production from biomass residue gasification in the state of Minas Gerais in Brazil. International Journal of Hydrogen Energy. 2025;**101**:358-378

[19] Ba Z, Zhao J, Li C, Huang J, Fang Y, Zhang L, et al. Developing efficient gasification technology for high-sulfur petroleum coke to hydrogen-rich syngas production. Fuel. 2020;**267**:117170

[20] Kiris B, Yasar M. Pet coke gasification via novel gasifier design: Experimental and machine learning based modeling approaches. International Journal of Hydrogen Energy. 2025;**140**:452-463

[21] Luo P, Chen Z, Chen X, Ma W. Deep desulfurization of high-sulfur petroleum coke via alkali catalytic roasting combined with ultrasonic oxidation. Materials. 2024;**17**:2609. DOI: 10.3390/ma17112609

[22] Davoodi A, Pakshir M, Babaiee M, Ebrahimi GR. A comparative H_2S corrosion study of 304L and 316L stainless steels in acidic media. Corrosion Science. 2011;**53**:399-408

[23] Hesketh J, Dickinson EJF, Martin ML, Hinds G, Turnbull A. Influence of H_2S on the pitting corrosion of 316L stainless steel in oilfield brine. Corrosion Science. 2021;**182**:109265

[24] Zhang L, Tang X, Wang Z, Li T, Zhang Z, Lu M. The corrosion behavior of 316L stainless steel in H_2S environment at high temperatures. International Journal of Electrochemical Science. 2017;**12**:8806-8819. DOI: 10.20964/2017.09.17

[25] Srinivas G, Gebhard S, Schwab S, Looker M. Poison-resistant water gas shift catalyst for biomass and coal gasification. In: 2019 Crosscutting Research, Gasification Research, DE-SC0004378. TDA Research Inc.; 2019

[26] Dou X, Veksha A, Chan WP, Oh WD, Liang YN, Teoh F, et al. Poisoning effects of H_2S and HCl on the naphthalene steam reforming and water-gas shift activities of Ni and Fe catalysts. Fuel. 2019;**241**:1008-1018

[27] Ray N, Rastogi VK, Mahapatra H, Sen SP. Deactivation of low temperature shift catalysts; part I poisoning by hydrogen sulphide. Journal of the Research Institute for Catalysis. 1973;**21**(3):187-199

[28] Méndez-Mateos D, Barrio VL, Requies JM, Cambra JF. A study of deactivation by H_2S and regeneration of a Ni catalyst supported on Al_2O_3, during methanation of CO_2. Effect of

the promoters Co, Cr, Fe and Mo. RSC Advances. 2020;**10**:16551-16564

[29] Sala L, Zaryab SA, Chiesa P, Martelli E. Comparison and optimization of CO_2 purification units for CCS applications. The International Journal of Greenhouse Gas Control. 2024;**136**:104193

Chapter 3

Hydrogen Production via Sustainable Gasification

Kunlanan Wiranarongkorn and Karittha Im-orb

Abstract

In order to move toward sustainability, many countries have made the production of renewable energy one of their main priorities due to the issues of global warming brought on by greenhouse gas (GHG) emissions and the decline in the supply of fossil fuels. Therefore, the H_2 production via sustainable processes is one option that can be done to relieve the problems. The chapter involves H_2 production via sustainable gasification using the renewable feedstock that is firstly discussed. Then the improvement of the gasification process to be more efficient, including the process of integration and the installation of the carbon capture and utilization system, is explained. Next, the environmental benefits in terms of carbon neutrality, waste reduction, and energy security of the sustainable gasification are presented. Finally, the future outlook of H_2 production via sustainable gasification is discussed.

Keywords: carbon neutrality, sustainability, biomass, gasification, hydrogen

1. Introduction

The world has started to recognize and prioritize the production of energy from alternative fuel sources, which are clean fuels with zero net carbon dioxide emissions, due to the serious issue of global climate change brought on by the high levels of GHG released when fossil fuels are burned for energy production. The global CO_2 emission has been found to continuously rise; the CO_2 level in the year 2022 increased by 1.5% relative to 2021 [1]. A number of nations decided to achieve worldwide net zero emissions by 2050 in order to alleviate this issue [2]. Many nations have implemented policies to encourage a rise in the share of fuel and energy generation from renewable sources in order to reach the global sustainability goals. The use of hydrogen (H_2) for energy production is a potential alternative to achieve the goal of zero GHG emissions, as there is no carbon dioxide emission from the process when H_2 is converted to energy through combustion or fuel cells. Furthermore, H_2 is an energy carrier that can be employed as a form of energy storage to accommodate the fluctuations in energy demand over time. However, more than 90% of the current H_2 production sources are produced from fossil fuels that still emit carbon dioxide from the process [3]. Therefore, in order to obtain clean energy, it is necessary to increase the proportion of green H_2 production, which is H_2 produced from renewable energy sources, for the creation of fuel, heat, and electricity.

IntechOpen

One clean and renewable energy source is biomass. The primary benefit of biomass energy is that the process of producing it does not raise the net quantity of carbon dioxide in the environment since the carbon dioxide generated during combustion is recycled into raw materials for plant photosynthesis. Biomass consists of a variety of biological materials whose main components are cellulose, hemicellulose, lignin, small amounts of lipids, proteins, sugars, starches, inorganic components, and water [4]. The various forms of biomass can be categorized according to their sources: (1) forestry crops and residues, (2) agricultural crops and residues, (3) industrial residues, (4) animal residues, (5) municipal solid waste (MSW), and (6) sewage. However, agricultural and wood products, biogas and landfill gas, MSW, and alcohol fuels like ethanol and biodiesel are the biomass products that are now being used in practice [5].

Gasification is a widely utilized method to turn solid biomass into fuel gas due to its high efficiency and can work with a variety of feedstocks. The generated gas can be cleaned to create syngas, which is mostly composed of H_2 and carbon monoxide, and high-purity H_2, or it can be utilized straight as fuel for a combustion unit [6]. Different biomass containing different elements provided different gasification performance. Moreover, the adjustment of process parameters could improve the H_2 yield. For example, He et al. [7] did an experiment on a fixed bed gasifier to compare the process performance when using different feedstock: citrus peel and pine sawdust. The results showed that the process using citrus peel provided a higher H_2 yield of 34.35 mol/kg biomass and carbon conversion efficiency of 66.30% compared to the process using pine sawdust. The technical viability of co-gasifying biomass and paper-mill sludge (PMS) for the production of renewable fuel was examined by Rosha et al. [8]. Although the synergy of biomass and PMS could improve H_2 content, the lower heating value (LHV) of product gas gradually decreased due to the increased moisture content. The gasification performance was also affected by the operating parameters. The H_2 content increased as gasifying temperature increased, whereas an opposite effect was found when ER increased. Lan et al. [9] studied the fluidized bed gasifier using pine wood. The results indicated that an increase in gasifying temperature could increase the calorific value and yield of the generated gas while decreasing the concentration of CO_2. However, an opposite trend was observed when ER increased.

Renewable H_2 can be produced via a biomass gasification (BG) process integrated with the cleaning and conditioning process, such as a water gas shift (WGS) unit, in which the CO reacts with water to form CO_2 and H_2, or it can be combined with cutting-edge CO_2 capture technologies (like sorption-enhanced biomass gasification) to produce negative carbon emissions. Several studies have investigated various aspects of the H_2 production process driven by BG in order to overcome technical and economic limitations and compete with H_2-based fossil fuel. Xu et al. [10] studied the BG integrated with an O_2 transport membrane and an H_2 separation membrane for H_2 production using artificial intelligence. The multi-objective optimization results revealed that the proposed process offered promising results, with a potential 26% reduction in power consumption. Furthermore, a 20% reduction in the payback period of the system was achieved, along with an 8% decrease in the levelized cost of H_2. Zhang et al. [11] investigated how various parameters affected the gasification of biomass in a downdraft gasifier coupled with a reforming H_2 generation using the process model. It was discovered that the product gas produced during the gasification and reforming operations had a high H_2 concentration of 98–99%. As global trends shift toward sustainability, sustainable H_2 production via gasification is receiving increasingly more attention. To overcome this challenge, several options can be done to achieve sustainable and carbon-neutral goals, such as the process efficiency improvement by adjusting the operating parameters, designing the new process configurations, or the reduction of CO_2 emissions by installing the CO_2 capture and utilization technologies. Therefore, this chapter critically reviews and analyzes H_2 production via the sustainable gasification process. The effect of operating parameters, including the type of biomass, gasifying agents, and gasifying conditions, on H_2 production is discussed. Then, the upgrading of the gasification process to meet the sustainable and carbon-neutral goals is explored. Finally, the future prospects of H_2 production via sustainable gasification are described.

2. H_2 production through gasification

Gasification is the thermochemical process in which carbon-based fuels (i.e., coal, biomass) react with gasifying agents such as the controlled amount of O_2 and/or steam to produce syngas, which consists of CO, H_2, CO_2, CH_4, tar, and char. Equations (1)-(9) display the primary reactions in the gasification process [12]. Subsequently, the produced gas is cleaned to remove some impurities. Then, the WGS reaction (Eq. (3)) transforms the CO in syngas into H_2. Finally, H_2 can be separated from the process using CO_2 capture technologies. **Figure 1** shows the simplified process flow diagram for the gasification-based H_2 production process.

$$\text{Water gas}: C + H_2O \leftrightarrow H_2 + CO + 131.5\, kJ/mol \tag{1}$$

$$\text{Methanation}: C + 2H_2 \leftrightarrow CH_4 - 74.8\, kJ/mol \tag{2}$$

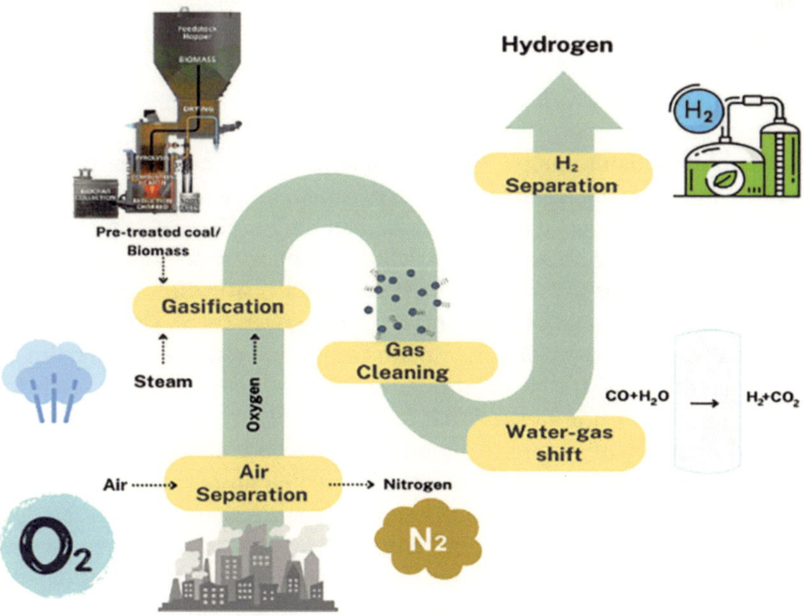

Figure 1.
Simplified process flow diagram of the H_2 production process via gasification.

$$\text{Water} - \text{gas shift}: CO + H_2O \leftrightarrow H_2 + CO_2 - 41\, kJ/mol \quad (3)$$

$$\text{Boudouard}: C + CO_2 \leftrightarrow 2CO + 172\, kJ/mol \quad (4)$$

$$\text{Methane dry reforming}: CH_4 + CO_2 \leftrightarrow 2H_2 + 2CO + 216\, kJ/mol \quad (5)$$

$$\text{Methane steam reforming}: CH_4 + H_2O \leftrightarrow 3H_2 + CO + 206\, kJ/mol \quad (6)$$

$$\text{Combustion}: H_2 + 1/2O_2 \leftrightarrow H_2O - 242\, kJ/mol \quad (7)$$

$$\text{Combustion}: CO + 1/2O_2 \leftrightarrow CO_2 - 283\, kJ/mol \quad (8)$$

$$\text{Combustion}: C + 1/2O_2 \leftrightarrow CO - 111\, kJ/mol \quad (9)$$

3. Operating parameters influencing gasification performance

To achieve the sustainable goal, biomass is an outstanding renewable feedstock for H_2 production through the gasification process due to its carbon-neutral characteristic. Several operating parameters of biomass gasification, such as feed characteristics, gasifying agents, and operating conditions such as gasifying temperature and pressure, are the main contributors that influence the gasification performance [13]. Detailed discussions of the main influences are shown in the following sections.

3.1 Characteristic of renewable feedstock

Different biomass containing different properties provided different gasification performance. Mu et al. [14] conducted an experiment on an oxygen-blown steam gasification of different lignocellulosic biomass, including soybean straw, rice straw, corn straw, wheat straw, cotton straw, maple sawdust, bamboo wood, and pine sawdust. The results indicated that soybean straw offered good gasification characteristics and provided the highest H_2 yield of 28.96 mol/kg. Darmawana et al. [15] investigated H_2 production from a mixture of empty palm fronds and waste-derived palm oil production plant through the gasification process using water at critical condition as a gasifying agent. The finding showed that the highest H_2 production efficiency of 73% was obtained when the mixing ratio of empty palm fronds and waste-derived palm oil production plant was 3. Kalinci et al. [16] studied H_2 production from different biomass, that is, almond shells, walnut branches, rice straw, wood chips, sludge, and wastepaper, using a combined gasification and power generation process. It was found that the H_2 concentration obtained from almond shells and wastepaper ranged from 7 to 18% by volume. Furthermore, the kind of biomass used had an impact on the process's energy and exergy efficiency; processes that used biomass with a high nitrogen content had lower energy and exergy efficiency. Atnaw et al. [17] found that using palm fronds as feedstock for the gasification process could produce syngas with a heating value similar to using wood chips. **Table 1** outlines the comparison of specific properties of different biomass feedstocks, which significantly influence H_2 yield.

The moisture content of biomass also influences the properties of produced syngas, efficiency, and energy requirement of the gasification process. Schuster et al. [22] found that more moisture in the feedstock accelerated the production of combustible gas during the gasification process. An increase in moisture content promotes the H_2 production due to the vital role of the WGS reaction. Motta et al. [23] informed that H_2 generation via fluidized bed gasifier could enhance from 34.69 to 49.42 vol% when the moisture content of biomass increased from 10 to 40%. As feedstocks with high moisture contents caused the reactor's temperature to decrease and slowed down the rate of endothermic reactions, 15 wt% water contents are usually recommended for most biomass sources.

The particle size of biomass is another important parameter affecting the gasification performance because it influences the heat and mass transfer circumstances. As the small particle biomass has a high surface area, it could well contact the gasifying agent and enhance the rate of gasification reactions [24]. Hernandez et al. [25] carried out an experiment on the gasification of biomass using an entrained-flow gasifier. The impact of biomass particle size was examined, and the findings showed that when particle size decreased from 8 mm to 0.5 mm, the H_2 concentration rose from 3.1% to 9.3%. The effect of biomass particle size on a downdraft fixed-bed gasifier's gasification performance was assessed by Tinaut et al. [18]. The results showed that using a smaller particle size and a lower air velocity offered the best efficiency.

3.2 Effect of gasifying agent

In biomass gasification, the biomass reacts with a restricted amount of gasifying agent to produce syngas. The gasifying agents that can be used in gasification are air, oxygen, steam, carbon dioxide, or a mixture of them. The different gasifying agents provide syngas with different H_2 concentrations and calorific values. Shayan et al. [26] investigated the use of different gasifying agents including, steam, air, oxygen-enriched air, and oxygen. It was discovered that the highest rate of H_2 production was reached when the steam was used as a gasifying agent, followed by O_2, O_2-enriched air, and air systems, respectively. Additionally, the highest thermal efficiency and exergy efficiency were achieved in air and steam systems. Habibollahzade et al. [27] investigated the influence of gasifying agents, those were, CO_2, steam, O_2, O_2-enriched air, air, and a mixture of them, on the process efficiency. The finding showed that using an O_2-based agent in the process provided higher cold gas efficiencies (CGE) when high carbon content biomass was used as feedstock, while the process using CO_2 agent offered higher CGE for the low carbon content biomass. Moreover, the gasification process using an O_2-based agent offered higher exergy efficiency. The effects of gasifying agents (air-steam, O_2-enriched air, O_2-steam, and air,) on sawdust gasification at the equivalent ratio (ER) of 0.2 to 0.3 were examined by Meng et al. [28]. The process with oxygen-enriched air provided the producer gas with a high LHV due to lower N_2 content; air-steam promoted the rate of H_2 production due to an increase in the extent of the WGS reaction. The gasification of aquatic biomass using CO_2 and O_2 as gasifying agents was studied by Hanaoka et al. [29]. The H_2 concentration of the syngas reduced as the CO_2 feed rate increased, whereas the CO concentration increased. However, when the O_2 feed rate grew, so did the H_2 and CO concentrations. A CO_2/O_2 ratio of 45/55 vol.% produced the highest syngas output of 69.7 vol.%. The range of H_2 generation for various gasifying agents is displayed in **Table 2**.

Parameters	Rice straw [18]	Wheat straw [19]	Cedarwood [20]	Rice husk [20]	Biomass with 30% plastic waste [21]
Ultimate analysis (% wt., dry and ash free basis)					
C	43.08	39.39	48.8	40.3	83.93
H	6.63	5.36	6.6	5.4	12.64
O	38.56	38.14	43	54.1	0.80
S	0.21	0.29	0.2	—	0.2
N	0.65	0.87	1.4	0.2	—
Proximate analysis (% wt., dry basis)					
FC	12.30	18.01	N/A	N/A	0.41
VM	76.84	66.04	N/A	N/A	96.88
Ash	10.86	15.95	N/A	N/A	2.43
Gasification technology	Chemical looping steam gasification of biomass with Fe_2O_3/CaO as the oxygen carrier	Two-stage steam gasification using a fixed bed gasifier, with biochar supported Ni-based catalyst	Steam gasification	Steam gasification	Autothermal downdraft co-gasification
Gasifying temperature (°C)	900	900	850	850	1017
H_2 content (vol%)	64.10	50.12	68.8	66.8	40.9
H_2 yield / g biomass	34.23 mmol	22.96 mmol	31.0 mmol	28.0 mmol	—

Table 1.
Properties of biomass feedstocks for H_2 production.

Gasifying agents	H_2 concentration (vol%)	H_2 yield (ml/g)	Ref.
Air	5–25	0–20	[30]
Steam	20–60	200–1400	[31]
CO_2/O_2	18–24	150–400	[32]

Table 2.
H_2 production from gasification using various gasifying agents.

3.3 Effect of gasifying conditions

Operating parameters influencing H_2 production and tar formation included gasifying temperature, gasifying pressure, and resident time. Increasing temperature increased the rate of biomass pyrolysis and cracking of volatile components, resulting in the improvement of the syngas production performance. A small-scale fluidized bed experiment was conducted by Cao et al. [33] to gasify pine sawdust using olivine as an in-bed material. By raising the gasifying temperature from 700 to 850°C, the concentration of CO and H_2 rose by 15.5% and 11%, respectively. The highest H_2 production of 71.4 g/kg and the lowest tar concentration of 1.3 g/Nm^3 were achieved at a gasifying temperature of 800°C, an Folivine content of 50%, and S/B of 1.2. Nguyen et al. [31] used torrefied wood chips as a fuel in a pilot-scale bubbling fluidized bed reactor to assess the H_2 generation via steam gasification. The results expressed that higher gasifying temperature and S/B were more conducive to H_2 generation and improved carbon conversion efficiency, while increased ER reduced yield and concentration of H_2. The optimal S/B at 850°C was 1.2, offering an H_2 yield of 0.039 Nm^3/h and an H_2 concentration of 48.41%. Li et al. [34] observed the effect of gasifying temperature on the H_2 production during BG at high temperature in a range of 800–1435°C. The highest H_2 production was achieved at a gasifying temperature of 917°C. Du et al. [35] reported that when the gasifying temperature rose from 1800 to 2000 K, the quantity of H_2 and CO molecules increased from 586 to 165 and 614 to 199, respectively. Moreover, Liu et al. [36] investigated the production of H_2-rich gas via BG using the process model developed in Aspen plus. The highest H_2 concentration of 14.9 vol% was attained when reaction temperature and S/B were 1123 K and 1.0, respectively.

The operating pressure is another parameter that influences the gasification performance. Yang et al. [37] investigated the production of H_2 during BG at various initial pressures (IP). At 1800 K and 1 bar, the number of CO molecules (n(CO)) and H_2 molecules (n(H_2)) in the basic structure were examined. Moreover, a molecular dynamics model was used to assess how variations in IP affected the synthesis of n(CO) and n(H_2) using 10% vol of a Pt catalyst. The n(CO) and n(H_2) increased from 165 to 586 and 175 to 598, respectively, when the IP increased from 1 to 5 bar. Renganathan et al. [38] found that the performance of CO_2 gasification was negatively impacted by the pressure. The CO content declined when gasifying pressure increased, whereas the H_2 content did not change.

4. Novel gasification technology for H_2 production

Numerous innovative gasification technologies including chemical looping gasification (CLG), sorption-enhanced gasification (SEG), and solar-driven gasification have been proposed to enhance process efficiency and mitigate environmental

Gasification technologies	Advantages	Disadvantages
CLG	• Reduce CO_2 emissions • An oxygen carrier can be a catalyst that promotes the gas production. • Reduce exergy loss	An oxygen carrier has low reactivity, low stability, and high cost.
SEG	• Enhance H_2 production • Reduce CO_2 emissions • Reduce generated tar	The catalyst and CO_2 sorbent have low reactivity, low cyclic stability, and poor absorption performance.
Solar-driven gasification	• Low CO_2 emissions • High energy efficiency • High heating rate	The equipment is complex.

Table 3.
Benefits and drawbacks of innovative biomass gasification technology [13].

impact. Nevertheless, the research and development stages of these gasification systems are still ongoing. Different BG technologies have offered different advantages and disadvantages, as summarized in **Table 3**.

4.1 Chemical looping gasification

Chemical looping gasification (CLG) of biomass is the technology for in-situ CO_2 capture developed based on the principle of the chemical looping combustion (CLC). The CLC consists of two reactors (fuel and air reactors), as shown in **Figure 2(a)**. The biomass reacts with lattice oxygen provided by the oxygen carrier, a solid substance that contains oxygen (usually metal oxides Me_xO_y) in the fuel reactor. Then the oxygen-depleted solid material is reoxidized by air in the air reactor before it starts a new cycle [39]. A typical CLG unit consists of a fuel reactor and an air reactor. The air reactor is utilized for oxygen carrier regeneration and residual char combustion, while the fuel reactor is used for BG [40], as illustrated in **Figure 2(b)**. When compared to alternative BG processes, the CLG of biomass offers a number of benefits, such as: (1) the high-temperature oxygen carrier from the air reactor can supply the thermal energy required for the endothermic reactions during gasification, (2) the oxygen carrier lowers the cost of pure oxygen and steam because it provides the lattice oxygen for gasification, and (3) the lattice oxygen is more prone to partially oxidizing the fuels compared with gas-phase oxygen, thus the syngas with a high LHV is obtained [41].

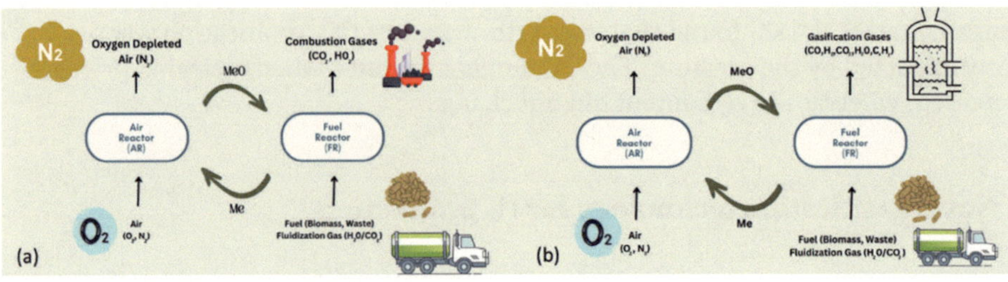

Figure 2.
Schematic diagrams of (a) CLC and (b) CLG processes.

Several works studied the H_2 production from CLG to improve the process performance. Different oxygen carriers were investigated to find the suitable one. Active transition metal oxides including NiO, CuO, Mn_2O_3, and Fe_2O_3 were typically the oxygen carriers, whereas TiO_2, Al_2O_3, SiO_2, bentonite, $MgAl_2O_4$, and ZrO_2 were the support materials. However, the iron oxides (Fe_2O_3, Fe_3O_4) were widely used due to their good stability, cheap price, and abundance. Liu et al. [42] studied CLG of microalgae using Fe_2O_3 as an oxygen carrier. The results indicated that the gasification efficiency could be improved from 61.65% to 81.64% due to the partial oxidation. And the Fe_2O_3 preserved good robustness after 10 cycles. Although the Fe_2O_3 offered several advantages, it is less active compared to nickel-based and copper-based material. Therefore, alkali and alkaline earth metals (AAEM), such as K, Na, and Ca, were introduced to increase the reactivity of iron-based oxygen carriers. Hu et al. [43] reported that using $Ca_2Fe_2O_5$ as an oxygen carrier provided H_2 production of 516.77 ml/g biomass at 800°C. However, this oxygen carrier could be used in a few cycles because the structure of Fe-Ca was easily destroyed. Lui et al. [44] used NiO/Al_2O_3 as a dual-functional material (a catalyst and an oxygen carrier) to study the CLG of waste plastic (polypropylene (PP)). The NiO/Al_2O_3 was robust and revealed the highest lattice oxygen utilization ratio.

The process configuration improvement of CLG of biomass was also investigated in several aspects. Zeng et al. [45] proposed high-quality syngas generation through the chemical looping pyrolysis-gasification (CLPG) of biomass, in which the fuel reactor was separated into two sections: (1) pyrolysis and (2) gasification. The CLPG process was an appropriate method for a solid fuel with high volatile contents. The results showed that the fuel reactor temperature of 820°C and S/B of 1.00 produced the best conditions, providing H_2/CO of 2.45 and CGE of 61.87%. Fu et al. [46] investigated the use of CLG in conjunction with microwave-assisted heating to produce H_2-rich gas from plastic trash. According to the results, a H_2 conversion efficiency of 84.3% and an H_2 yield of 46.0 mmol/g were achieved. Gua et al. [47] proposed a biomass CLG combined with CO_2 absorption in a dual circulating fluidized bed (DCFB). The proposed CLG process revealed: (1) an enhanced heat transfer coefficient of particles, (2) an improved gas-solid mixing, (3) a 23.71% decrease in CO_2 content, (4) a 10.11% increase in H_2 content, and (5) the particles were densely distributed in the upper part of the reactor. Lin et al. [48] investigated the complex gas-solid interactions in the CLG systems as well as the optimization of reactor setup. The finding indicated that increasing the gasifier height enhanced H_2 production and fuel conversion but compromised CO_2 absorption.

4.2 Sorption-enhanced gasification

Sorption-enhanced steam biomass gasification (SEG) (**Figure 3**) is a new technique for producing H_2-rich gas that combines steam biomass gasification with calcium looping. This procedure involves adding calcium oxide (CaO) to the gasifier, which acts as both a CO_2 sorbent and a tar-cracking catalyst. In a gasifier, CO_2 generated during steam gasification is captured by CaO and converted to calcium carbonate ($CaCO_3$) via a carbonation reaction. The equilibrium of the WGS reaction (Eq. (3)) will be shifted forward by the in-situ removal of CO_2, and as a result, the H_2 production is improved. The heat required to decompose $CaCO_3$ and regenerate CaO is produced in the combustor by char combustion [49].

Although significant progress has been reported in the literatures, there are still a lot of uncertainties regarding the SEG, especially in relation to process performance

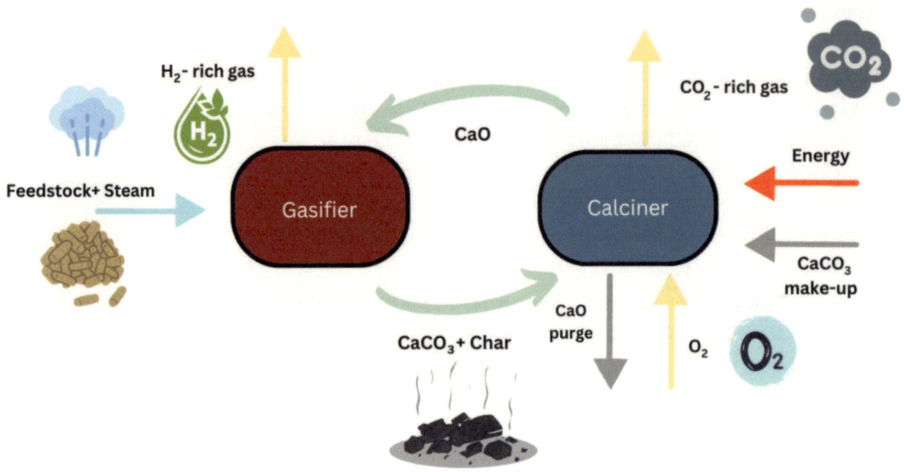

Figure 3.
Schematic diagram of the SEG process.

for H_2 production under relevant conditions. In order to proceed with the catalytic gasification of biomass and sorption-enhanced reforming procedures, Li et al. [50] investigated a modified two-stage SEG for H_2 production utilizing a dual fixed-bed system. This involved separating the gasifier of the standard SEG method into two independent reactors. While the sorbent/carbon ratio was 0.82, the H_2 concentration and yield could reach 70.5 vol% and 600.7 mL/g-SB, respectively. Cormos [51] reported that green H_2 generation using biomass SEG provided appealing results in terms of a high overall energy efficiency of approximately 50%, specific carbon emissions below 7 kg/MWh, and negative CO_2 emissions of 518.40 kg/MWh for the entire biomass value chain.

The changes in operating parameters also influence the SEG performance. The impact of the S/B on the efficiency of biomass SEG for H_2 production was investigated by Inayat et al. [52]. According to the results, the H_2 concentration at S/B of 3.5 was approximately 12% greater than that at 1.0. Kalinci et al. [53] stated that the increase in steam promoted the decomposition of tar and coke and enhanced H_2 production via steam methane reforming (SMR) and WGS reactions. de Lasa et al. [54] reported that H_2 production and heating value increased noticeably when gasification temperature increased, whereas tar and coke concentration significantly decreased.

4.3 Solar-driven gasification

Although the CLG and SEG present a potential alternative, they still rely on fossil fuels for energy production units, leading to extensive carbon emissions. The use of a solar-driven biomass gasifier, which integrated biomass gasification with solar heating, instead of fossil fuels to provide the required energy, is an interesting option. The solar-driven gasification offers several benefits, such as reduction in energy input, reduction in CO_2 emissions, and improvement of energy efficiency. Pozzobon et al. [55] used a xenon arc lamp to simulate sunlight in order to study solar-driven gasification. The sample's surface temperature reached 1500°C. According to the data, using woodchip with 50% moisture content as feedstock produced syngas with 8.5 vol% CO_2, 13 vol% CH_4, 31 vol% CO, and 38 vol% H_2. A direct irradiation solar gasification reactor with an average solar flux of 1171.3 kW/m^2 with a maximum reaction bed temperature of 1260°C was designed by Wang et al. [56]. The cellulose entirely decomposed, and at an energy upgrade factor of 1.15, syngas with a 47.1% H_2 content was produced. Deshmukh and Santhosh [57] investigated the green H_2 production process through a solar-assisted gasification of agricultural residues. Response surface methodology (RSM) was used to optimize the process. The results showed that the producer gas with a 40% H_2 concentration, a higher heating value (HHV) of 6 MJ/kg, and a CGE of 80% was achieved when the ER was 0.2, the gasifying temperature was between 800 and 900°C, and the steam-to-air mass ratio (S/A) was between 1.8 and 2.0. Bai et al. [58] reported that the solar-driven BG offered higher H_2/CO (1.43–1.89) compared to traditional biomass gasification.

5. Environmental benefits

In this section, the environmental aspects of H_2 production from gasification systems will be proposed, including processes featuring novel gasifier technologies, waste feedstock utilization, downstream process improvements, and carbon capture and utilization (CCU). Many researchers have evaluated emissions of GHGs, especially CO_2, from this process via life cycle assessment (LCA) to identify the hot spot and suggest key improvements for the development of sustainable gasification technology. Hamedani et al. [59] investigated the environmental performance of 1 MW H_2 production from indirect heated biomass gasification using almond shell biomass. They found that the global warming potential (GWP) of this process was 0.042 kg CO_2 per 1 MJ of H_2, with the main contribution to GWP arising from syngas production via biomass gasification (34% of total GWP). This was attributed to flue gas emissions from the gasifier combustors. It was notable that the process also generated indirect GHG emissions from raw material sources such as the production of gasifying agents (steam, O_2, and CO_2) and the heat sources required to supply the gasifier and downstream processes. Consequently, some researchers have explored the use of solar radiation as a renewable heat source for biomass gasification [60, 61]. Takeda et al. [60] proposed the use of circulating fluidized bed gasification of forest residues for H_2 production, utilizing indirect heating from solar energy. Their process offered a GWP of 1.04 kg CO_2-eq/kg H_2, which was 50% less than that of conventional gasification (2.67 kg CO_2-eq/kg H_2). Similarly, Chen et al. [61] conducted a LCA of supercritical water gasification of biomass utilizing solar energy. They found that the potential of this solar-powered gasification process as an environmentally beneficial technique was demonstrated by their finding that its GWP was 4.41 kg CO_2-eq/kg H_2.

Some studies evaluate the environmental performance of processes incorporating novel gasifier technologies, either to handle biomass with high moisture content or to improve efficiency. For example, Peng et al. [62], Qi et al. [63], and Hu et al. [64] employed supercritical water gasification technology to convert oilfield wastewater, organic wastewater, and sewage sludge, respectively, into H_2. In Peng et al. [62]'s work, the process utilizing heat recovery and carbon capture and storage (CCS) resulted in a GWP of 5.52 kg CO_2 eq/kg H_2, with the H_2 separation section being the primary contributor to GWP. Meanwhile, Qi et al. [63] reported a minimum GWP of 5.27 kg CO_2 eq/kg H_2 for their gasification system integrated with a multi-effect evaporator and CCS. In Hu et al. [64]'s study, the GWP of their proposed process with CCS was 0.31 kg CO_2 eq/kWh. Additionally, Wu et al. [65] presented biomass gasification integrated with chemical looping H_2 production. In the chemical looping process, syngas from biomass gasification was fed into fuel reactors for reduction reactions, while air was fed into air reactors for oxidation reactions. This system produced purified H_2 and captured CO_2 while achieving negative CO_2 emissions of −15.13 and − 17.00 kg CO_2 eq/kg H_2 when air and oxygen were used as gasifying agents, respectively. However, chemical looping H_2 production was the largest contributor to GWP, accounting for 39.03% of the total.

For sustainable feedstock in gasification system, using waste materials such as plastic waste, MSW, organic waste, and so on is an efficient method to reduce CO_2 emissions from the overall process. Al-Rowaili et al. [66] studied the oxygen gasification of vacuum residue, incorporating H_2S removal, WGS, and CO_2 capture subunits for high-purity H_2 production. Additionally, dry reforming of natural gas using CO_2 reactants was integrated to produce more synthesis gas. They discovered that the overall CO_2 emissions were 8.3 kg CO_2/kg H_2. Qi et al. [67] presented a process for H_2 production from MSW via plasma gasification (PG) integrated with chemical looping. In this system, syngas from the plasma gasifier was fed into a chemical looping unit consisting of fuel, steam, and air reactors to simultaneously produce H_2 and capture CO_2. A comparison between their proposed process and conventional biomass gasification revealed that the conventional process exhibited a 93.76% higher GWP, primarily due to high CO_2 emissions from PSA and acid gas removal units. However, the PG was the main source of GWP, contributing 70.05% and 83.98% of the total GWP in the conventional and novel procedures, respectively. Moreover, Afzal et al. [68] presented a mixed plastic waste (MPW) indirect heated gasification process. Their findings indicated that MPW gasification was more environmentally beneficial than the conventional gasification process, with a GWP of 10.8 kg CO_2 eq/kg H_2.

Based on the studies mentioned above, it can be concluded that the main source of GHG emissions from conventional biomass gasification processes lies in downstream units, such as acid gas removal and PSA units. These units require significant energy input to remove undesired gases, such as H_2S and CO_2, as well as unreacted fuel, which contributes to CO_2 emissions [67]. Consequently, several studies have proposed improvements in purification technologies to reduce CO_2 emissions in biomass gasification processes.

Process intensification technologies have been employed to reduce the number of purification steps. Tang et al. [69] proposed a sorption-enhanced water-gas shift (SEWGS) integrated with PG for high-purity H_2 production from refuse-derived fuel (RDF) obtained from excavated waste, as shown in **Figure 4**. The GWP of the entire life cycle of the PG-SEWGS process was 645 kg CO_2-eq/ton RDF, which was 16.88% lower than that of conventional biomass gasification, which included Rectisol acid gas removal (AGR). The main subsystem contributing to GHG emissions was the biomass pretreatment stage, including excavated waste mining to RDF. Additionally, CCS technology was directly integrated into the gasification system to further reduce CO_2 emissions, as reported by Hren et al. [70]. Their environmental analysis revealed that the gasification process combined with CCS could achieve negative CO_2 emissions when operating under electricity self-sufficient conditions, with values ranging from −9.56 to −18.8 kg CO_2-eq/kg H_2.

The integration of biomass gasification with CCU has been extensively studied to enhance CO_2 emission reduction performance. Researchers have explored integrated CO_2 capture technologies, such as monoethanolamine (MEA) absorption, adsorption, chemical looping, and SEWGS technologies, to improve environmental performance (for example, see [62, 63, 67, 69, 71].). However, CO_2 is also released during the CO_2

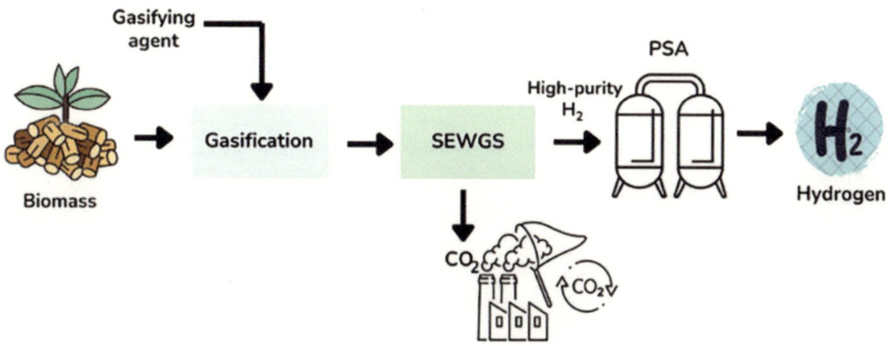

Figure 4.
Biomass gasification combines with SEWGS for high purity H_2 production.

capture and storage processes. To address this, various integrated biomass gasification processes coupling CO_2 utilization have been proposed.

For methanol and DME synthesis from biomass gasification, these processes require a suitable syngas ratio of approximately 2 for methanol production and 1 for DME, necessitating complex syngas conditioning steps, including tar removal, WGS reactions, and AGR (as shown in **Figure 5**). These downstream processes significantly contribute to CO_2 emissions. Consequently, researchers have proposed novel designs for the syngas conditioning stage or advanced purification technologies to reduce the number of downstream steps and enhance separation efficiency.

For instance, Liu et al. [72] proposed integrating biomass gasification and methanol synthesis by incorporating SEWGS technology in the syngas cleaning stage to simultaneously produce more H_2 and capture CO_2. Additionally, they applied the reverse Boudouard (RB) reaction to produce tunable syngas from CO_2 for methanol synthesis. Their results showed that the GWP of biomass gasification to methanol production at optimal conditions was 2857.18 kg CO_2-eq/ton biomass, with biomass gasification contributing 92% of the total GWP across the entire process, including collection and transportation of biomass and transportation of CO_2 and methanol. Liu et al. [73] further compared the environmental performance of two CO_2 utilization methods in integrated biomass gasification and methanol synthesis: the RB and dry reforming reactions. Their results indicated that the RB reaction yielded a lower

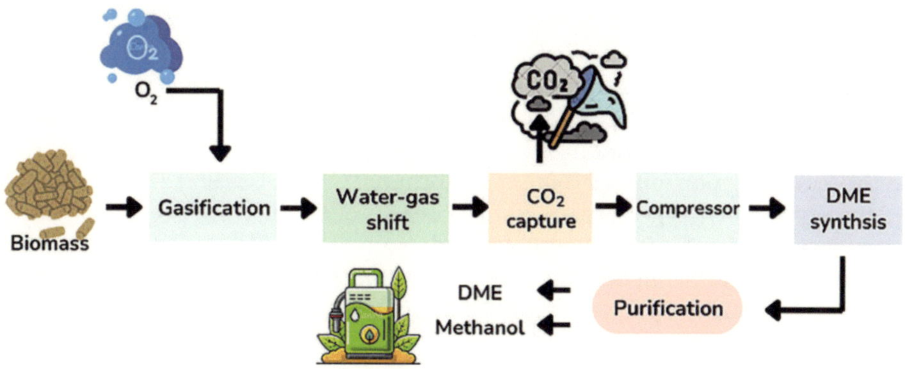

Figure 5.
Biomass gasification with CCU through methanol and DME synthesis.

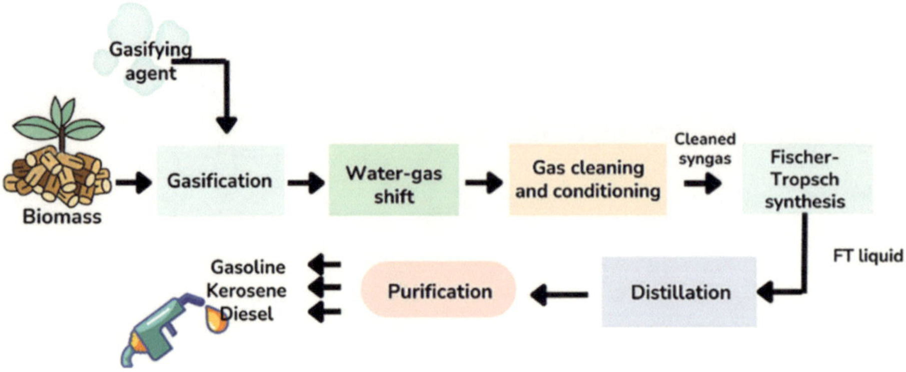

Figure 6.
Biomass gasification with CCU through multi-product production.

GWP than dry reforming reaction due to reduced exhaust gas emissions in the pressure gasification unit (PGU).

In DME synthesis, Im-orb and Piroonlerkgul [74] designed an integrated biomass gasification and DME synthesis process using oil palm empty fruit bunch (EFB). Their results showed that biomass gasification was the primary contributor to GWP, accounting for 91% of the total carbon footprint. Yadav et al. [75] compared the environmental performance of two syngas cleaning methods: WGS-AGR and WGS-Zeolite membrane. Their study found that the biomass gasification process using WGS-membrane technology for syngas cleaning prior to methanol synthesis achieved a GWP of 48.2 kg CO_2-eq/ton methanol, 23.61% lower than the process using WGS-AGR. This reduction was due to higher CO_2, CO, and sulfur emissions from the AGR process compared to the membrane technology.

Figure 6 illustrates the process flow diagram of biomass gasification integrated with Fischer-Tropsch (FT) synthesis to produce H_2, FT liquid fuel, and electricity. Raw syngas from the biomass gasification section is fed into the gas conditioning unit, which includes a WGS reactor and an AGR. The cleaned syngas is then introduced into the FT synthesis reactor to produce FT liquid fuel, which is subsequently sent to a distillation unit to separate products such as kerosene, gasoline, and diesel. In this integrated process, residual fuel and heating and cooling energy can be recycled back into the system. Consequently, a number of researchers have proposed improving the biomass-to-FT liquid fuel processes by integrating innovative technology to recover leftover fuel and energy, thereby reducing the environmental impact.

For example, Jasper et al. [76] recovered flue gas from the FT synthesis unit to produce additional H_2 via an autothermal reformer and electricity through a combustor and gas turbine. The GWP of process was negative at −0.1355 kg CO_2/GJ when 24% of the off-gas was fed into a combined heat and power (CHP) unit to generate electricity, and the remaining portion was routed to reforming. García-Casas et al. [77] evaluated the environmental performance of kerosene production from oxygen-blown gasification integrated with FT synthesis and upgrading. Additionally, autothermal reforming of flue gas from the PSA unit was incorporated to produce more syngas. From an environmental perspective, this process had a 47.96% efficiency and a GWP of 0.049 kg CO_2/kWh.

The GWP comparison of several configurations in gasification with different biomass feedstock is shown in **Table 4**.

Biomass	Process configuration	GWP ($kg\ CO_2\ eq/kg\ H_2$)	References
Agro-Industrial Residue	Air indirect gasification/WGS/PSA/CHP system	0.042 ($kg\ CO_2eq/MJ\ H_2$)	[59]
Biomass	Solar-driven indirect gasification (CFB)/ cyclone separator/WGS/PSA	1.04	[60]
Wood sawdust	Supercritical water gasification solar system/ cyclone separator/WGS/PSA without CCS and with CCS	4.41 and – 11.06	[61]
Oilfield wastewater	Supercritical water gasification/Oxidation unit/ gas-liquid separator/PSA/CCS	5.52	[62]
Organic wastewater	Supercritical water air gasification/Oxidation unit/gas-liquid separator/PSA/CCS	5.27	[63]
Rice straw	BG/chemical looping H_2 production/waste heat recovery/CCS	–15.13 (air gasification	[64]
Sewage sludge	Supercritical water gasification/ oxidation unit/ CHP system/ PSA for H_2/ CCS	0.31	[65]
Vacuum Residue	Gasification/H_2S removal/ dry methane reforming/WGS/CO_2 capture	8.3	[66]
MSW	PG/ WGS/AGR/PSA	631.49 CO_2-eq/ton biomass	[67]
	PG/ synthetic chemical looping	261.84 CO_2-eq/ton biomass	
Plastic waste	CLG /H_2S removal/steam reformer/WGS/PSA	10.8	[68]
Excavated waste	PG/SEWGS/AGR	645 kg CO_2-eq/ton RDF	[69]
Biomass	Steam gasification/tar reforming/scrubbing/ H_2S removal/WGS/PSA	–9.56 to –18.8 kg CO_2-eq/kg H_2.	[70]
Sawdust	Air separation unit/O_2-CO_2 gasification/ H_2S removal/ organic Rankine cycle unit	97.69 kg CO_2-eq/MWh	[71]
Corncob	Steam gasification/ SEWGS/ RB reaction	2857.18 kg CO_2-eq/ton biomass	[72]
Pine stem wood	Pre-treatment/entrained flow BG/WGS/ membrane/MeOH synthesis	48.2 kg CO_2-eq/ton methanol	[75]
Raw pine sawdust	O_2/steam gasification/WGS/AGR/ FT synthesis/ATR/CHP	–0.1355 kg CO_2/GJ	[76]
	BG/gas cleaning/WGS/AGR/FT synthesis and upgrading	0.049 kg CO_2/kWh,	[77]

Table 4.
GWP of sustainable gasification for H_2 production.

6. Future outlook

The performance of BG has been studied and developed across various aspects. Many studies have sought to improve the BG efficiency by investigating the effect of operating conditions, proposing novel gasification and separation technologies, and introducing new process designs. However, the efficiency remains relatively low (40–50%), which negatively impacts both production cost and environmental performance. To address this, additional methods for improving efficiency should be explored. Examples include material recovery techniques, such as recycling unreacted gas, designing multiproduct generation processes, and implementing power consumption reduction strategies. These strategies could involve using high efficiency technologies and applying heat integration designs.

From an economic perspective, the gasifier and gas cleaning and conditioning systems have been improved to reduce capital investment as well as raw material and utility costs. Nevertheless, the prices of final products such as H_2, methanol, and FT liquid remain high due to the costs of raw materials and energy. Consequently, processes utilizing cheaper raw materials such as biomass residue and MSW through improved renewable energy efficiency should be further studied.

The integrated process of BG with carbon capture, storage, and utilization by converting CO_2 into high-value chemicals is a promising approach to reducing GHG emissions. However, reducing indirect GHG emitted from utility and raw material sources in downstream processes, such as acid gas removal, CO_2 capture, and product purification, remains a significant challenge. To address this, novel process designs should focus on using reactants and utilities produced from renewable sources, such as using O_2 and H_2 generated through electrolysis and energy derived from solar and wind power, to enhance environmental benefits. Additionally, there are limited studies addressing the trade-off between environmental and economic performance, as well as between environmental and technical performance. Further investigation into these topics could provide valuable guidance for making informed decisions about suitable conditions and configuration routes for the integrated gasification process.

7. Conclusions

The sustainable H_2 production via gasification is discussed, and the biomass is recognized as a potential feedstock for agricultural countries. Different biomass with different characteristics in terms of fixed carbon, volatile matter, moisture content, ash, and constituent elements provides the syngas with different H_2 concentrations and calorific values. Although using biomass with a high moisture content resulted in a high H_2 concentration, the gasification process requires a significant amount of energy. The enhancement of the H_2 production and process efficiency of the gasification process can be done by adjusting operating parameters and developing novel gasification technologies, that is, CLG, SEG, and solar-driven gasification. The CLG and SEG can improve H_2 production while decreasing CO_2 emissions; however, they are still in the research and development stages in which the catalyst and CO_2 sorbent with high reactivity and good stability are required to improve. The solar-driven gasification can decrease the energy input, improve energy efficiency, and decrease the CO_2 emission, but there are limitations to the complexity of the process. In terms of sustainability, the negative GHG emissions represents a significant milestone for the gasification process. Utilizing waste as feedstock, absorbing and utilizing emitted CO_2, improving downstream process efficiency, and implementing advanced gasification technology into action are some possible approaches. However, as the suggested sustainable gasification method for producing H_2 is still in the research and development phase, more study from a technological, financial, and environmental perspective should be carried out on a regular basis.

Acknowledgements

This work was supported by School of Food Industry, King Mongkut's institute of Technology Ladkrabang, and Department of Basic Science and Physical Education, Faculty of Science at Sriracha, Kasetsart University.

Author details

Kunlanan Wiranarongkorn[1] and Karittha Im-orb[2*]

1 Department of Basic Science and Physical Education, Faculty of Science at Sriracha, Kasetsart University, Chonburi, Thailand

2 Program in Food Process Engineering, School of Food-Industry, King Mongkut's Institute of Technology Ladkrabang, Bangkok, Thailand

*Address all correspondence to: karittha.im@kmitl.ac.th

IntechOpen

© 2025 The Author(s). Licensee IntechOpen. This chapter is distributed under the terms of the Creative Commons Attribution License (http://creativecommons.org/licenses/by/4.0), which permits unrestricted use, distribution, and reproduction in any medium, provided the original work is properly cited.

References

[1] Lui Z, Deng Z, Davis S, Ciais P. Monitoring global carbon emissions in 2022. Nature Reviews Earth & Environment. 2023;**4**:205-206. DOI: 10.1038/s43017-023-00406-z

[2] UN Climate change conference 2021. COP26 Explained [Internet]. 2021. Available from: https://ukcop26.org/ [Accessed: September 29, 2022]

[3] U.S. Department of Energy. Hydrogen Strategy Enabling a Low-Carbon Economy. Vol. 22022. Available from: https://www.energy.gov/fecm/downloads/hydrogen-strategy-enabling-low-carbon-economy [Accessed: October 29, 2022]

[4] Pandey B, Prajapati YK, Sheth PN. Recent progress in thermochemical techniques to produce hydrogen gas from biomass: A state of the art review. International Journal of Hydrogen Energy. 2019;**44**:25384-25415. DOI: 10.1016/j.ijhydene.2019.08.031

[5] Dincer I, Ishaq H. Renewable Hydrogen Production. Oxford: Elsevier; 2022. 364 p. DOI: 10.1016/C2020-0-02435-7

[6] Swain PK, Das LM, Naik SN. Biomass to liquid: A prospective challenge to research and development in 21st century. Renewable and Sustainable Energy Reviews. 2011;**15**(9):4917-4933. DOI: 10.1016/j.rser.2011.07.061

[7] He J, Yang Z, Guo M, Gu L, Zhang L, Yan Y, et al. Effect of gasifying agents on sawdust gasification in a novel pilot scale bubbling fluidized bed system. International Journal of Hydrogen Energy. 2022;**2022**(47):30428-30439. DOI: 10.1016/j.fuel.2019.03.107

[8] Rosha PJ, Ibrahim H. Technical feasibility of biomass and paper-mill sludge co-gasification for renewable fuel production using Aspen plus. Energy. 2022;**258**:24883. DOI: 10.1016/j.energy.2022.124883

[9] Lan W, Chen G, Zhu X, Wang X, Wang X, Zu B. Research on characteristics of biomass gasification in a fluidized bed. Journal of the Energy Institute. 2019;**92**:613-620. DOI: 10.1016/j.joei.2018.03.011

[10] Xu G, Qian H, Zhang Q, Alsennani TR, Bouzgarrou S, Alturise F. Integration of biomass gasification and O_2/H_2 separation membranes for H_2 production/separation with inherent CO_2 capture: Techno-economic evaluation and artificial neural network based multi-objective optimization. Renewable Energy. 2024;**224**:20150. DOI: 10.1016/j.renene.2024.120150

[11] Zhang Y, Zhou A, Li Z, Zhang H, Xiong Y, Xiao R, et al. Numerical simulation analysis of biomass gasification and rich-H_2 production process in a downdraft gasifier. Journal of the Energy Institute. 2024;**114**:01596. DOI: 10.1016/j.joei.2024.101596

[12] Doherty W, Reynolds A, Kennedy D. The effect of air preheating in a biomass CFB gasifier using Aspen plus simulation. Biomass and Bioenergy. 2009;**33**:1158-1167. DOI: 10.1016/j.biombioe.2009.05.004

[13] Sikarwar VS, Zhao M, Clough P, Yao J, Zhong X, Memon MZ, et al. An overview of advances in biomass gasification. Energy and Environment Science. 2016;**9**:2939-2977. DOI: 10.1039/C6EE00935B

[14] Mu Q, Aleem RD, Liu C, Elendu CC, Cao C, Duan PG. Oxygen

blown steam gasification of different kinds of lignocellulosic biomass for the production of hydrogen-rich syngas. Renewable Energy. 2024;**232**:121132. DOI: 10.1016/j.renene.2024.121132

[15] Darmawana A, Ajiwibowo MW, Cahyo FA, Aziz M, Tokimatsu K. Co-production of hydrogen and power from palm mill wastes. Energy Procedia. 2019;**158**:1891-1896. DOI: 10.1016/j.egypro.2019.01.437

[16] Kalinci Y, Hepbasli A, Dincer I. Efficiency assessment of an integrated gasifier/ boiler system for hydrogen production with different biomass types. International Journal of Hydrogen Energy. 2010;**35**:4991-5000. DOI: 10.1016/j.ijhydene.2009.08.079

[17] Atnaw SM, Sulaiman SA, Yusup S. Syngas production from downdraft gasification of oil palm fronds. Energy. 2013;**61**:491-501. DOI: 10.1016/j.energy.2013.09.039

[18] Tinaut FV, Melgar A, Perez JF, Horrillo A. Effect of biomass particle size and air superficial velocity on the gasification process in a downdraft fixed bed gasifier. An experimental and modelling study. Fuel Processing Technology. 2008;**89**(11):1076-1089. DOI: 10.1016/j.fuproc.2008.04.010

[19] Yao D, Hu Q, Wang D, Yang H, Wu C, Wang X. Hydrogen production from biomass gasification using biochar as a catalyst/support. Bioresource Technology. 2016;**216**:159-164. DOI: 10.1016/j.biortech.2016.05.011

[20] Anniwaer A, Chaihad N, Zhang M, Wang C, Yu T, Kasai Y. Hydrogen-rich gas production from steam co-gasification of banana peel with agricultural residues and woody biomass. Waste Management. 2021;**125**:204-214. DOI: 10.1016/j.wasman.2021.02.042

[21] Ranjan N, Yadav N, Singh H, Kumar S, Mahajani SM. Modelling and simulation of autothermal downdraft co-gasification of biomass and plastic wastes using Aspen plus. Energy Conversion and Management. 2023;**297**:117714. DOI: 10.1016/j.enconman.2023.117714

[22] Schuster G, Loffler G, Weigl K, Hofbauer H. Biomass steam gasification—An extensive parametric modeling study. Bioresource Technology. 2001;**77**:71-79. DOI: 10.1016/S0960-8524(00)00115-2

[23] Motta IL, Miranda NT, Maciel Filho R, Wolf Maciel MR. Biomass gasification in fluidized beds: A review of biomass moisture content and operating pressure effects. Renewable and Sustainable Energy Reviews. 2018;**94**:998-1023. DOI: 10.1016/j.rser.2018.06.042

[24] Luo S, Xiao B, Guo X, Hu Z, Liu S, He M. Hydrogen-rich gas from catalytic steam gasification of biomass in a fixed bed reactor: Influence of particle size on gasification performance. International Journal of Hydrogen Energy. 2009;**34**:1260-1264. DOI: 10.1016/j.ijhydene.2008.10.088

[25] Hernandez JJ, Aranda-Almansa G, Bula A. Gasification of biomass wastes in an entrained flow gasifier: Effect of the particle size and the residence time. Fuel Processing Technology. 2010;**91**:681-692. DOI: 10.1016/j.fuproc.2010.01.018

[26] Shayan E, Zare V, Mirzaee I. Hydrogen production from biomass gasification; a theoretical comparison of using different gasification agents. Energy Conversion and Management. 2018;**159**:30-41. DOI: 10.1016/j.enconman.2017.12.096

[27] Habibollahzade A, Ahmadi P, Rosen MA. Biomass gasification using

various gasification agents: Optimum feedstock selection, detailed numerical analyses and tri-objective grey wolf optimization. Journal of Cleaner Production. 2021;**284**:124718. DOI: 10.1016/j.jclepro.2020.124718

[28] Meng F, Ma Q, Wang H, Liu Y, Wang D. Effect of gasifying agents on sawdust gasification in a novel pilot scale bubbling fluidized bed system. Fuel. 2019;**249**:112-118. DOI: 10.1016/j.fuel.2019.03.107

[29] Hanaoka T, Hiasa S, Edashige Y. Syngas production by CO_2/O_2 gasification of aquatic biomass. Fuel Processing Technology. 2013;**116**:9-15. DOI: 10.1016/j.fuproc.2013.03.049

[30] Cao Y, Bai Y, Du J. Air-gasification of pine sawdust using dolomite as in-bed material: Effect of gasification conditions on product characteristics. Journal of the Energy Institute. 2021;**95**:187-192. DOI: 10.1016/j.joei.2021.01.010

[31] Nguyen NM, Alobaid F, May J, Peter J, Epple B. Experimental study on steam gasification of torrefied woodchips in a bubbling fluidized bed reactor. Energy. 2020;**202**:117744. DOI: 10.1016/j.energy.2020.117744

[32] Flori G, Frigo S, Barontini F, Gabrielli R, Sica P. Experimental assessment of oxy-CO_2 gasification strategy with woody biomass. Renewable Energy. 2024;**228**:120593. DOI: 10.1016/j.energy.2020.117744

[33] Cao Y, Bai Y, Du J. Study on gasification characteristics of pine sawdust using olivine as in-bed material for combustible gas production. Journal of the Energy Institute. 2021;**96**:168-172. DOI: 10.1016/j.joei.2021.03.014

[34] Li W, Li Q, Chen R, Wu Y, Zhang Y. Investigation of hydrogen production using wood pellets gasification with steam at high temperature over 800°C to 1435°C. International Journal of Hydrogen Energy. 2014;**39**:5580-5588. DOI: 10.1016/j.ijhydene.2014.01.102

[35] Du C, Nasajpour-Esfahani N, Hekmatifar M, Toghraie D, Esmaeili S. The effect of initial temperature and external heat flux on the H_2 and CO production by biomass gasification using molecular dynamics simulation. Renewable Energy. 2023;**215**(1):119011. DOI: 10.1016/j.renene.2023.119011

[36] Liu Z, Zhao C, Cai L, Long X. Steady state modelling of steam-gasification of biomass for H_2-rich syngas production. Energy. 2022;**238**:121616. DOI: 10.1016/j.energy.2021.121616

[37] Yang Y, Al-dolaimy F, Sajadi SM, Hussein Wais AM, Sabetvand R, Abbas AHR, et al. Molecular dynamics simulation of gasification technology to produce hydrogen from biomass at different initial pressures in the presence of platinum catalyst. International Journal of Hydrogen Energy. 2024;**56**(10):7-15. DOI: 10.1016/j.ijhydene.2023.11.304

[38] Renganathan T, Yadav MV, Pushpavanam S, Voolapalli RK, Cho YS. CO_2 utilization for gasification of carbonaceous feedstocks: A thermodynamic analysis. Chemical Engineering Science. 2012;**83**:159-170. DOI: 10.1016/j.ces.2012.04.024

[39] Lv L, Zhang Z, Li H. SNG-electricity cogeneration through MSW gasification integrated with a dual chemical looping process. Journal Chemical Engineering and Processing: Process Intensification. 2019;**145**:107665. DOI: 10.1016/j.cep.2019.107665

[40] Nguyen NM, Alobaid F, Dieringer P, Epple B. Biomass-based chemical looping

gasification: Overview and recent developments. Applied Sciences. 2021;**11**(15):7069. DOI: 10.3390/app11157069

[41] Ge H, Guo W, Shen L, Song T, Xiao J. Biomass gasification using chemical looping in a 25 kWth reactor with natural hematite as oxygen carrier. Chemical Engineering Journal. 2016;**286**:174-183. DOI: 10.1016/j.cej.2015.10.092

[42] Liu G, Liao Y, Wu Y, Ma X, Chen L. Characteristics of microalgae gasification through chemical looping in the presence of steam. International Association for Hydrogen Energy. 2017;**42**(36):22730-22742. DOI: 10.1016/j.ijhydene.2017.07.173

[43] Hu Q, Shen Y, Chew JW, Ge T, Wang CH. Chemical looping gasification of biomass with Fe_2O_3/CaO as the oxygen carrier for hydrogen-enriched syngas production. Chemical Engineering Journal. 2020;**379**:122346. DOI: 10.1016/j.cej.2019.122346

[44] Liu B, You C, Wang H. Chemical looping gasification of waste plastics for syngas production using NiO/Al_2O_3 as dual functional material. Chemical Engineering Journal. 2024;**501**:157728. DOI: 10.1016/j.cej.2024.157728

[45] Zeng J, Xiao R, Zhang H, Wang Y, Zeng D, Ma Z. Chemical looping pyrolysis-gasification of biomass for high H_2/CO syngas production. Fuel Processing Technology. 2017;**168**:116-122. DOI: 10.1016/j.fuproc.2017.08.036

[46] Fu W, Zhang Y, Cao W, Zhao W, Li B. Microwave-assisted chemical looping gasification of plastics for H_2-rich gas production. Chemical Engineering Journal. 2024;**499**:156225. DOI: 10.1016/j.cej.2024.156225

[47] Guo M, Lin J, Yu J, Wang S, Luo K, Fan J. Configuration optimization of a biomass chemical looping gasification (CLG) system combined with CO_2 absorption. Renewable Energy. 2024;**237**:121459. DOI: 10.1016/j.renene.2024.121459

[48] Lin J, Luo K, Yu J, Wang S, Fan J. Numerical simulation of gasifier optimization for combined hydrogen production and carbon reduction in a chemical looping gasification (CLG) system. Chemical Engineering Science. 2025;**301**:120694. DOI: 10.1016/j.ces.2024.120694

[49] Doranehgard MH, Samadyar H, Mesbah M, Haratipour P, Samiezade S. High-purity hydrogen production with in situ CO_2 capture based on biomass gasification. Fuel. 2017;**202**:29-35. DOI: 10.1016/j.fuel.2017.04.014

[50] Li C, Gond X, Zhang H, Zhang Y. A modified two-stage sorption-enhanced steam gasification of biomass process for H_2 production. Fuel. 2023;**352**:129018. DOI: 10.1016/j.fuel.2023.129018

[51] Cosmos CC. Decarbonized green hydrogen production by sorption-enhanced biomass gasification: An integrated techno-economic and environmental evaluation. International Journal of Hydrogen Energy. 2024;**95**:592-603. DOI: 10.1016/j.ijhydene.2024.11.281

[52] Inayat A, Ahmad MM, Abdul Mutalib MI, Yusup S. Effect of process parameters on hydrogen production and efficiency in biomass gasification using modelling approach. Journal of Applied Sciences. 2010;**24**(10):3183-3190. DOI: 10.3923/jas.2010.3183.3190

[53] Kalinci Y, Hepbasli A, Dincer I. Biomass-based hydrogen production: A review and analysis. International Journal of Hydrogen Energy. 2009;**34**(21):8799-8817. DOI: 10.1016/j.ijhydene.2009.08.078

[54] de Lasa H, Salaices E, Mazumder J, Lucky R. Catalytic steam gasification of biomass: Catalysts, thermodynamics and kinetics. Chemical Reviews. 2011;**111**(9):5404-5433. DOI: 10.1021/cr200024w

[55] Pozzobon V, Salvador S, Bezian JJ. Biomass gasification under high solar heat flux: Experiments on thermally thick samples. Fuel. 2016;**174**:257-266. DOI: 10.1016/j.fuel.2016.02.003

[56] Wang S, Zhu X, Liu YZ, Bai Z, Liu Q, Huang X, et al. Design and experimental study of solar-driven biomass gasification based on direct irradiation solar thermochemical reactor. Chemical Engineering Journal. 2024;**500**:157062. DOI: 10.1016/j.cej.2024.157062

[57] Deshmukh S, Santhosh R. Solar-assisted gasification of agriculture residues for green hydrogen production. Bioresource Technology Reports. 2023;**22**:101506. DOI: 10.1016/j.biteb.2023.101506

[58] Bai Z, Liu Q, Lei J, Li H, Jin H. A polygeneration system for the methanol production and the power generation with the solar-biomass thermal gasification. Energy Conversion and Management. 2015;**102**:190-201. DOI: 10.1016/j.enconman.2015.02.031

[59] Hamedani SR, Villarini M, Colantoni A, Moretti M, Bocci E. Life cycle performance of hydrogen production via agro-industrial residue gasification—A small scale power plant study. Energies. 2018;**11**:675. DOI: 10.3390/en11030675

[60] Takeda S, Nam H, Chapman A. Low-carbon energy transition with the sun and forest: Solar-driven hydrogen production from biomass. International Journal of Hydrogen Energy. 2022;**47**:24651-24668. DOI: 10.1016/j.ijhydene.2021.11.203

[61] Chen J, Xu W, Zuo H, Wu X, Jiaqiang E, Wang T, et al. System development and environmental performance analysis of a solar driven supercritical water gasification pilot plant for hydrogen production using life cycle assessment approach. Energy Conversion and Management. 2019;**184**:60-73. DOI: 10.1016/j.enconman.2019.01.041

[62] Peng Z, Wang L, Yi L, Xu J, Liu Z, Jin H, et al. Performance assessment of an energetically self-sufficient system for hydrogen production from oilfield wastewater treated by supercritical water gasification. International Journal of Hydrogen Energy. 2024;**53**:907-918. DOI: 10.1016/j.ijhydene.2023.11.315

[63] Qi X, Ren Z, Meng F, Lu L, Liu F, Li X, et al. Thermodynamic and environmental analysis of an integrated multi-effect evaporation and organic wastewater supercritical water gasification system for hydrogen production. Applied Energy. 2024;**357**:122449. DOI: 10.1016/j.apenergy.2023.122449

[64] Hu D, Ren C, Zhang S, Ma M, Chen Y, Chen B, et al. Thermodynamic and environmental analysis of integrated supercritical water gasification of sewage sludge for power and hydrogen production. Energy. 2024;**299**:131568. DOI: 10.1016/j.energy.2024.131568

[65] Wu D, Gao Z, Wu S, Xiao R. Negative net global warming potential hydrogen production through biomass gasification combined with chemical looping: Environmental and economic assessments. International Journal of Hydrogen Energy. 2024;**66**:24-32. DOI: 10.1016/j.ijhydene.2024.04.078

[66] Al-Rowaili FN, Khalafalla SS, Jamal A, Al-Yami DS, Zahid U, Al-Mutairi EM. Techno-economic evaluation of hydrogen production via gasification of vacuum residue integrated with dry methane reforming. Sustainability. 2021;**13**:13588. DOI: 10.3390/su132413588

[67] Qi H, Cui P, Liu Z, Xu Z, Yao D, Wang Y, et al. Conceptual design and comprehensive analysis for novel municipal sludge gasification-based hydrogen production via plasma gasifier. Energy Conversion and Management. 2021;**245**:114635. DOI: 10.1016/j.enconman.2021.114635

[68] Afzal S, Singh A, Nicholson SR, Uekert T, DesVeaux JS, Tan ECD, et al. Techno-economic analysis and life cycle assessment of mixed plastic waste gasification for production of methanol and hydrogen. Green Chemistry. 2023;**25**:5068-5085. DOI: 10.1039/d3gc00679d

[69] Tang J, Tang Y, Liu Y, Sun Z, Deng J, Chen W, et al. From excavated waste to hydrogen: Life cycle techno-environmental-economic comparison between plasma-gasification-based water gas shift and sorption enhanced water gas shift routes. Energy Conversion and Management. 2023;**292**:117375. DOI: 10.1016/j.enconman.2023.117375

[70] Hren R, Vujanovic A, Fan YV, Klemeš JJ, Krajnc D, Čuček L. Hydrogen production, storage and transport for renewable energy and chemicals: An environmental footprint assessment. A Renewable and Sustainable Energy Reviews. 2023;**173**:113113. DOI: 10.1016/j.rser.2022.113113

[71] Wang Y, Zhu L, He Y, Zeng X, Hao Q, Huang Y, et al. Life cycle assessment of an efficient biomass power plant supported by semi-closed supercritical CO_2 cycle and chemical looping air separation. Science of the Total Environment. 2024;**919**:170832. DOI: 10.1016/j.scitotenv.2024.170832

[72] Liu H, Tang Y, Ma X, Tang J, Yue W. Biomass gasification based on sorption-enhanced hydrogen production coupled with carbon utilization to produce tunable syngas for methanol synthesis. Energy Conversion and Management. 2024;**309**:118428. DOI: 10.1016/j.enconman.2024.118428

[73] Liu H, Tang Y, Ma X, Tang J, Deng J, Yue W. Calcium looping-enhanced biomass gasification for methanol production: Integrating methane dry reforming and carbon utilization. Separation and Purication. Technology. 2025;**354**:129377. DOI: 10.1016/j.seppur.2024.129377

[74] Im-orb K, Piroonlerkgul P. Sustainability analysis of the bio-dimethyl ether (bio-DME) production via integrated biomass gasification and direct DME synthesis process. Renewable Energy. 2023;**208**:324-330. DOI: 10.1016/j.renene.2023.03.092

[75] Yadav P, Athanassiadis D, Yacout DMM, Tysklind M, Upadhyayula VKK. Environmental impact and environmental cost assessment of methanol production from wood biomass. Environmental Pollution. 2020;**265**(Part A):114990. DOI: 10.1016/j.envpol.2020.114990

[76] Jasper M, Rafati N, Schimmel K, Shahbazi A, Li F, Mba-Wright M, et al. Carbon negative transportation fuels—A techno-economic-environmental analysis of biomass pathways for transportation. Energy Conversion and Management. 2022;**14**:100208. DOI: 10.1016/j.ecmx.2022.100208

[77] García-Casas M, Gálvez-Martos JL, Dufour J. Environmental and economic multi-objective optimisation of synthetic fuels production via an integrated methodology based on process simulation. Computers and Chemical Engineering. 2022;**157**:107624. DOI: 10.1016/j.compchemeng.2021.107624

Section 3

Biomass and Wastes Gasification

Chapter 4

Biomass Gasification for Sustainable Energy Production: Effect of Operational Parameters on Product Gas

Tesfa Nega Gesese, Asmare Tezera Admase, Ejigayehu Desalegn Asrade and Eshetu Getahun

Abstract

The increasing demand for energy, reliance on fossil fuels, heightened environmental concerns, and the political commitments established in the Paris Climate Agreement drive the pursuit of new energy sources that are more sustainable and compatible with environmental protection. Biomass has emerged as a primary renewable energy resource, offering significant advantages in terms of its diversity, availability, and sustainability for meeting energy needs in heating, electricity generation, and biofuel production for transportation, among other applications. Various strategies have been explored for effectively utilizing biomass, ranging from biological to thermochemical conversion methods. Gasification is a thermochemical process recognized as one of the most effective methods for energy recovery from biomass, producing syngas primarily composed of hydrogen (H_2), carbon monoxide (CO), and methane (CH_4). Currently, various parameters influencing the yield of product gas and the performance of the gasifier have garnered significant attention from researchers. This paper aims to review the theory and process of biomass gasification, including the different types of gasifiers. It compiles key operational and performance parameters of the gasification process, as well as their influence on gasification conditions and products. This approach seeks to provide a comprehensive overview of hydrogen-rich syngas production based on current technologies and industrial/commercialization pathways.

Keywords: operational parameter, product gas, biomass, gasification, sustainability, energy production

1. Introduction

Global energy demand has steadily risen over the past few decades due to rapid industrialization and improved living standards. This increase has led to significant environmental damage, contributing to global warming and climate change [1].

The world is facing a crucial time for energy due to the consumption of fossil fuels (natural gas, coal, and oil), the rise in fuel prices, and unacceptable environmental effects in recent years [2, 3]. There is an urgent and growing need to develop clean and renewable energy sources to reach carbon neutrality by 2050 [4]. The energy supply will become a more significant issue as the population grows [5]. The rising uncertainty in fossil fuel prices and their impact on climate change have led to a growing global demand for clean energy, particularly as industrialization accelerates [6]. Greenhouse gas (GHG) emissions from human activities, particularly the burning of fossil fuels for power generation, are significant contributors to climate change [7]. This situation underscores the need to transition from conventional energy sources to renewables. Consequently, there is an increasing focus on renewable resources like solar, biomass, and wind for sustainable development. Biomass, a significant and historic source of energy, accounts for about 14% of global renewable energy use, with rural areas in developing countries relying on it for up to 90% of their energy needs [8]. As around 90% of the world's population is projected to live in developing countries by 2050, biomass is expected to remain a crucial energy source for these populations. The exploitation of biomass as an energy resource can provide dual benefits in terms of reducing carbon dioxide emissions and enhancing fuel security, as it is abundant [1, 9]. Biomass gasification is unique in that it exclusively uses renewable resources, offering various benefits related to its environmental impact and carbon-neutral properties [4]. Biomass has notable advantages over other renewable sources due to its versatility and accessibility. As of 2011, biomass accounted for over 10% of the global energy supply and is vital in rural areas of developing countries for cooking and heating, highlighting its importance for energy security and sustainability [10–12].

2. Gasification process

Gasification is a process that involves the indirect combustion of solid and liquid biomass, transforming these materials into combustible syngas. This method serves as an alternative to traditional combustion, offering the advantage of significantly reducing the emission of dust and toxic gases [13, 14]. Gasification is capable of processing a diverse array of biomass feedstock, including woody residues, agricultural byproducts, and dedicated energy crops, without requiring significant modifications to the fundamental process [15, 16]. Pyrolysis is the thermal conversion of organic materials in the absence of oxygen, typically at lower temperatures, resulting in liquids as the primary product, along with potential chemical and food byproducts [17]. In contrast, gasification involves the thermal conversion of organic materials at elevated temperatures and under reducing conditions, primarily producing permanent gases, with char, water, and condensates as secondary products; it can be categorized into partial oxidation and indirect heating (steam gasification). Combustion, on the other hand, is the thermal conversion of organic matter with an oxidant, usually oxytem, leading to the production of carbon dioxide and water, where the oxidant is present in stoichiometric excess for complete oxidation [18, 19]. Rapid climate change and the anticipated impacts of global warming have made access to clean and renewable energy essential for sustainable development worldwide. Biomass-based energy, one of the oldest sources of renewable energy, plays a crucial role in meeting daily energy needs. Among the various bioenergy technologies, biomass gasification stands out as a significant method for producing heat, power, and biofuels for practical applications. Biomass gasification is the process by which carbonaceous solid fuels originating from organic matter are thermally converted into a gaseous energy medium mainly made up of carbon monoxide (CO) and hydrogen (H_2). In biomass gasification, numerous intricate and interconnected thermochemical reactions are involved. Regardless of the type of gasifier, the process of gasifying biomass can be broadly divided into four stages: (i) drying the biomass particles; (ii) pyrolysis of the dried biomass particles, also known as devolatilization; (iii) partial oxidation of the char and/or pyrolysis gases; and (iv) char gasification, also known as reduction [20].

However, despite the extensive literature and advancements in technology and materials, the adoption of gasification technology still faces significant barriers to widespread acceptance compared to conventional energy sources [21, 22]. **Figure 1** depicts the CO_2 gasification reactivity of char from high-ash biomass. This process involves the addition of oxygen, air, steam, or a combination of these as an oxidant [24, 25]. The resulting producer gas also contains smaller amounts of carbon dioxide (CO_2), methane (CH_4), ethane (C_2H_6), water (H_2O), and nitrogen (N_2) when air is used as the oxidant, along with various contaminants such as carbon particulates, tar, ash, and higher hydrocarbons [22, 26]. On an atomic level, biomass materials are primarily composed of carbon, hydrogen, oxygen, and nitrogen [27]. Thermochemical and bio-chemical (biological) conversion processes are the two primary methods for utilizing biomass as an energy source. The thermochemical approach includes four main options: combustion, gasification, pyrolysis, and liquefaction, while the bio-chemical method encompasses digestion and fermentation. Among the various conversion processes for lignocellulosic biomass, gasification is considered a particularly promising technology due to its significant potential and the possibility of advanced applications [28–30]. **Table 1** illustrates the main steps of the gasification process.

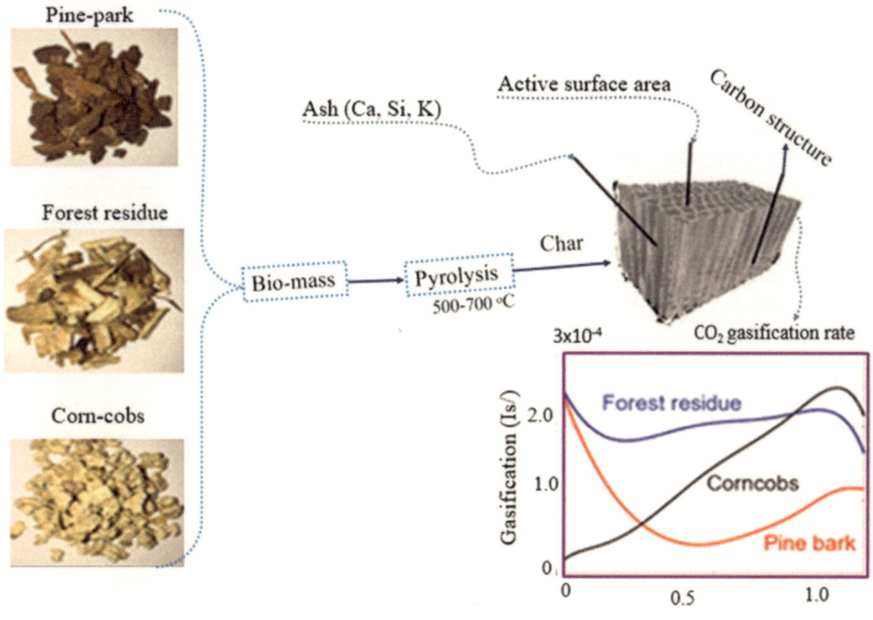

Figure 1.
CO_2 gasification reactivity of char from high-ash biomass [23].

Step	Rxn	Description
1. Oxidation	Exothermic	In this initial stage, biomass reacts with an oxidant (such as oxygen or air) to release heat and produce gases like CO and CO_2. This reaction generates the energy necessary for subsequent processes.
2. Drying	Endothermic	During drying, moisture is removed from the biomass feedstock. This step requires energy, typically provided by the heat generated in the oxidation stage, to reduce the moisture content to optimal levels.
3. Pyrolysis	Endothermic	In this stage, the biomass is subjected to heat in the absence of oxygen, leading to the thermal decomposition of organic materials. This process produces volatiles, including gases and tars, along with solid char.
4. Reduction	Endothermic	The reduction phase involves a series of reactions among the products of drying and pyrolysis, as well as with the gasifying agent. This step converts the solid char into combustible gases, such as hydrogen and methane.

Table 1.
The main steps of the gasification process [9].

2.1 Mechanisms of biomass gasification

Lignocellulosic biomass comprises varying amounts of cellulose, hemicellulose, lignin, and a small quantity of extractives, primarily made up of carbon, hydrogen, and oxygen. When used as feedstock in a gasifier, it undergoes a series of sequential steps: drying, pyrolysis (or devolatilization), reduction, and combustion. In the drying phase, moisture is removed, which is typically high in biomass. During pyrolysis, volatiles are released as gaseous CO, CO_2, and light hydrocarbons, along with liquid long-chain hydrocarbons known as tar, represented by the reaction:

$$CH_aO_b + O_2 + H_2O \rightarrow C + Tar + H_2 + CO + H_2O + CO_2 + CH_4$$

The yields of gas, liquid, and solid products depend on the properties of the feedstock, operating conditions, and the gasifying agent; however, not all liquid products can be completely converted into gases due to physical and chemical limitations, leading to tar contaminants in the final product gas. In the reduction phase, various chemical reactions occur among the products of drying and pyrolysis and the gasifying agents, while in the combustion phase, residual char is further combusted,

Reaction type	Reaction	Purpose
Partial combustion	$C + O_2 \rightarrow CO$	Produces syngas for further reactions
Combustion	$C + O_2 \rightarrow CO_2$	Generates heat and converts carbon to CO_2
Boudouard reaction	$C + CO_2 \rightarrow CO$	Enhances syngas yield from CO_2
Water-gas reaction	$C + H_2O \rightarrow CO + H_2$	Produces hydrogen for syngas enhancement
Hydrogasification	$C + H_2 \rightarrow CH_4$	Increases the calorific value of syngas
Water-gas shift	$CO + H_2O \rightarrow H_2 + CO_2$	Adjusts H_2/CO ratio for optimal use
Methanation	$CO + H_2 \rightarrow CH_4 + H_2O$	Increases energy content of gas

Table 2.
Key gasification reactions [33].

generating additional gaseous products and the necessary heat for the other processes. Generally, pyrolysis and combustion happen more rapidly than gasification, making gasification the rate-controlling step [27, 31–33]. In the biomass gasification process, many reactions take place at the same time and often compete with each other [34]. **Table 2** illustrates the most common gasification reactions in the process.

3. Gasifier type

The gasifier is a crucial component of the biomass gasification system, as it is the vessel in which all of the chemical reactions occur [35]. In a gasifier, carbonaceous fuel is subjected to various reactions and/or processes. Thus, a variety of gasifier types could be employed to complete the gasification process. Numerous types of gasifiers are employed for the gasification process; however, fixed bed gasifiers, fluidized bed gasifiers, entrained flow gasifiers, and plasma-type gasifiers are the most versatile gasifiers in the production of synthesis gas [36].

3.1 Fixed bed gasifier

The oldest type of gasifiers are the so-called fixed bed gasifiers, also called moving bed gasifiers. Based on the directions of the flow of the oxidant and the fuel, gasifiers can be classified as namely updraft gasifiers, downdraft gasifiers, and cross-draft gasifiers [37]. They have a limited capacity but are very simple to design and run [38]. These types of gasifiers are known to be used for small-scale applications. For this reason, fixed bed gasifiers are mostly utilized in decentralized biomass-based power generation [39]. The most widely used gasifiers in commercial settings are fixed bed models due to their straightforward construction and ease of usage. The fuel travels down the gasifier in the shape of a plug, hence the term "moving bed." The fixed bed gasifier operates in the pressure range of 0–70 bars. Fixed bed gasifiers have the following gas compositions: 3–5% CH_4, 10–15% CO_2, 10–15% CO, 15–20% H_2, and 40–50% N_2 [39].

3.1.1 Updraft or countercurrent gasifiers

This reactor converts biomass using drying, pyrolyzation, reduction, and hearth zones while using gas at the top and air at the bottom. It is easy to use, has a low gas-exit temperature, is very thermally efficient, and can gasify biomass with a high moisture content without the need for pre-drying. Furthermore, for this gasifier, size parameters are not very important. It does, however, generate a lot of tar [38]. These kinds of gasifiers produce a considerable amount of tar (about 50 g/Nm^3) because of the comparatively low gas temperature [40].

3.1.2 Downdraft or co-current gasifiers

The downdraft gasifier draws gas from the bottom after adding air to a packed bed of solid fuels that flow downhill. Although the arrangement is a little different, the zones are comparable to those in the updraft gasifier. Common issues for small downdraft gas producers include decreased overall efficiency and challenges managing increased moisture and ash content. Apart from these disadvantages, downdraft gasifiers must sustain consistently high temperatures across a specific cross-sectional area within the reaction chamber. Because of these issues, downdraft gasifiers can only be used at power levels below 1 MW [38]. Additionally, as compared to updraft gasifiers, downdraft gasifiers generate less tar. This is because the tar content generated during pyrolysis will undergo oxidation and disintegration into lighter substances [41].

3.1.3 Cross-draft gasifier

The top of the gasifier is where the fuel is introduced. One side of the gasifier introduces the gasifying chemical, while the other side removes the resulting gas. Nearly equal levels are maintained for the gasifying agent's entry and the resultant gas's exit. The fuel undergoes dryness, devolatilization, pyrolysis, and gasification as it passes through the gasifier before exiting. Near the gasifying agent's entrance is the oxidation zone, and close to the exit is the gasification zone. The oxidation/reduction zone is beneath the pyrolysis zone, while the drying zone is above the pyrolysis zone. The separation of the oxidation and reduction zones in the cross-draft gasifier ash bin restricts the use of various fuel types. The cross-draft gasifier's extremely high exit gas temperature affects the composition of the gas by producing gas with a higher carbon monoxide content and a lower hydrogen and methane content. Due to its compact nature, it could be preferable for small scale. However, it is not suitable for scale-up and reduced efficiency due to the exit gas's high temperature [37].

3.2 Fluidized bed gasifier

The fuel is injected into a hot sand bed that is either suspended (bubbling fluidized bed) or circulating (circulating fluidized bed). The bed is characterized by extreme turbulence and acts like a fluid [38]. The melting points of the bed material and ash determine the operating temperature of fluidized bed reactors, which is limited to 923–1223 K. The pressure is between 0 and 70 bar [39]. Although fluidized bed gasification is generally appropriate for large capacity (more than 10 M), it was first created to address operational issues with fixed bed gasification of fuels with high ash content. The consequent high tar content (up to 500 mg/Nm^3), incomplete carbon combustion, and poor response to load variations are the main issues with fluidized bed gasification. Certain biofuels may have issues with feeding, reaction bed instability, and fly-ash sintering in the gas channels. Fluidized bed gasifiers come in two main varieties: circulating fluidized bed (CFB) and bubbling fluidized bed (BFB) [38].

Fluidized bed gasifiers have several advantages over fixed bed gasifiers, including good gas-solid contact, better temperature control, good heat transfer characteristics, high volumetric capacity [40] a high carbon conversion rate, low tar production, a uniform and predictable temperature due to efficient gas-solid mixing, and flexibility in terms of fuel types, input rate, particle size, and moisture content. Reaction rates in fluidized bed gasifiers are significantly faster than in fixed bed gasifiers due to close gas-solid contact and the larger solid surface area brought about by lower particle sizes. The aforementioned characteristics make operating and scaling up fluidized bed gasifiers considerably simpler. Fluidized bed gasifiers could solve the majority of the issues with fixed bed gasifiers. Thus, compared to fixed bed gasifiers, fluidized bed gasifiers are more widely used [37]. Furthermore, high-pressure loss and corrosion are two drawbacks of fluidized reactors, which should be built to stop leaks. These can be divided into three categories: circulating, bubbling, and spouted fluidized beds [40].

3.3 Entrained flow gasifier

The entrained flow gasifier requires extremely small fuel particles and gasifies them using a high-temperature, turbulent flow. Although air or oxygen can be used as a gasifying agent, oxygen is used in the majority of commercial plants. High throughput and quick fuel conversion are made possible by the gasifier's high operating temperatures. The majority of the ash is eliminated as slag, and the resultant gas has less tar and condensable gases. Any fuel can be gasified with this gasifier; however, low moisture and ash content are preferred for lowering oxygen consumption. In addition, this kind of gasifier could be employed for large-scale production (> 100 MW), and it is efficient. Entrained flow gasification is a thoroughly studied process that is used in large-scale integrated gasification combined cycle (IGCC) coal power plants to gasify fossil fuels including coal and refinery waste. Nevertheless, its use in biomass gasification is currently being developed [37]. Almost 100% carbon conversion efficiency is attained by an entrained flow reactor gasifier running at 20–70 bar of pressure, which makes it appropriate for large-scale applications and tar cracking at temperatures higher than 1273 K [39].

3.4 Plasma gasifier

Plasma gasification is a high-temperature waste conversion process that generates synthesis gas, a valuable combustible gas, from the input material. Operating at temperatures ranging from 3500 to 6500°C, the reaction harnesses the thermal energy of the plasma and takes place within the reactor shell [42]. Fuel is introduced from the top of a gasifier and passes through the plasma zone as it is heated by a plasma system. Under the influence of the plasma system, gasification occurs at extremely high temperatures. The gasifying agent breaks down fuel into its component elements and transforms organic molecules into fuel gas by acting as an ionized gas. The top is cleared of gas, and any remaining inorganic material melts to form vitrified slag. Although it has advantages in reducing waste volume, being flexible in gasifying safely hazardous and toxic materials, and producing stable slag, it requires high capital costs and a high energy-consuming process [37].

4. Effect of parameters on the gasification process

The gasification process is significantly influenced by various operating parameters [1, 43–45]: (1) Equivalence ratio: The ratio of gasifying agent to fuel particles should be between 0.33 and 0.63 for optimal effectiveness; (2) Moisture content: Affects the pyrolysis stage by decreasing its reaction rate and impacting the quality of the end product; (3) Gas flow velocity: Low velocities result in slow pyrolysis conditions, leading to increased formation of solid and non-converted tars instead of lighter gaseous products; (4) Temperature: Higher temperatures and heating rates enhance tar reforming and carbon conversion efficiency, improving the quantity and quality of producer gas; (5) Atmosphere: The type of gasifying agent or mixture directly affects the syngas composition; (6) Residence time: Influences the yields and composition of the three gasification end products (condensable and non-condensable gases and char) and the thermal cracking of heavy tars; (7) feedstock physical and chemical properties; and (8) use of catalysts. Overall gasification performance, or its ability to convert solid material and energy content into combustible gas, can be assessed through specific parameters [46–48]: (1)

lower heating value (LHV) of the gas (MJ/Nm^3), which considers the gas's chemical composition and heating value; (2) carbon conversion efficiency (CCE, %), the ratio of carbon leaving with syngas to carbon entering the gasifier; (3) cold gas efficiency (CGE, %), the ratio of energy flow in the gas to energy contained in the biomass; and (4) H_2/CO volume ratio, which indicates processing requirements for end-use products [49]. For example, a hydrogen to carbon monoxide ratio greater than 1 indicates syngas suitable for chemical industries, such as methanol and ammonia, due to the high hydrogen and nitrogen content. Conversely, a ratio below 1 suggests the syngas is more appropriate for energy production in engines or gas turbines, as a ratio below 1 is better for energy production in engines or gas turbines [50].

4.1 Steam to biomass ratio (S/B)

The steam to biomass (S/B) ratio indicates the feed of steam (g/min) per feed of biomass (g/min). The product gas quality is highly dependent upon the S/B ratio. It can affect the quality of product gas yields and requirements for energy input [51]. Inayat et al. [52] studied the impact of process parameters on hydrogen production and efficiency in biomass steam gasification with in-situ CO_2 capture. They found that both temperature and steam/biomass (S/B) ratio positively influence hydrogen yield. For example, at 800 K and a low S/B ratio of 2.0, the hydrogen yield was 78.5 g/kg of biomass. In contrast, at the same temperature with a higher S/B ratio of 5.0, the yield increased to 96 g/kg, resulting in a 17.5 g/kg increment. At a higher temperature of 1300 K with a low S/B ratio of 2.0, the yield was 88.5 g/kg, which increased to 97 g/kg with a higher S/B ratio of 5.0. The difference in hydrogen yield at higher temperatures was 8.5 g/kg, lower than the difference observed at lower temperatures, indicating that the steam feed rate has a more significant impact at lower temperatures. This is due to the endothermic forward water-gas shift reaction being favored at lower temperatures, which is highly dependent on the steam feed rate. Overall, as the S/B ratio increases, hydrogen production rises while the amounts of CO and CH_4 decrease. Sadhwani et al. [34] examine how temperature affects the yields of char, liquid, and gas from the process. **Table 3** shows that as temperature rises during CO_2 gasification, char yield decreases significantly, while gas yield increases. Cao et al. [53] reported that increasing the S/B ratio from 0.61 to 2.7 raised H_2 from 16.78 vol. % to 19.64 vol. %, demonstrating that the water-gas shift (WGS) reaction was enhanced. The WGS reaction is essential in steam gasification, as it significantly alters the gas composition with changes in reaction rates.

4.2 Effect of temperature

Temperature is a key factor in gasification, affecting devolatilization and product gas composition. Increased temperatures improve biomass conversion to gas due to

Product stream	Temperature (K)			
	700	790	850	934
Char, wt. %	34.28 ± 2.15	31.13 ± 2.3	15.6 ± 0.94	12.89 ± 0.8
Liquid, wt. %	14.29 ± 1.11	15.85 ± 0.3	9.87 ± 0.51	10.57 ± 0.1
Gas, wt. %	51.42 ± 1.1	53.02 ± 2.5	74.54 ± 0.4	76.53 ± 0.9

Table 3.
Product stream yield in air gasification [34].

its highly volatile matter [43]. Gasification temperature also influences product gas quality, tar formation, reactor needs, and capital costs [39].

Khezri et al. [54] examine how reactor temperature, equivalence ratio (ER), and static bed height (SBH) affect the gasification performance of Napier grass and the combustibility of producer gas using a small-scale auto-thermal bubbling fluidized gasifier. As shown in **Figure 2**, hydrogen yield increases from 5.53% at 625°C to 15.36% at 825°C with rising temperature. Carbon monoxide (CO) yield also rises from 7.27% at 625°C to a peak of 12.13% at 725°C but then declines slightly to 10.51% at 825°C. In contrast, carbon dioxide (CO_2) output decreases rapidly with increasing temperature, falling to 9.04% at 725°C and reaching a minimum of 4.14% at 825°C. At 725°C, methane (CH_4) concentration gradually decreases to 2.48%. This decline is likely due to the dominance of exothermic reactions, such as the water-gas shift and methanation reactions, which occur at lower temperatures, along with the low reactivity of endothermic reactions like the Boudouard reaction, water-gas reaction, and steam methane reforming.

Jamin et al. [55] reported that the composition of hydrogen gas increased from 6.33 to 17.22 vol. % for raw wood waste and from 6.17 to 16.43 vol. % for torrefied wood waste when the gasification temperature was raised from 600 to 1000°C. Concurrently, the carbon monoxide gas composition for both feedstocks also improved, rising from 11.86 to 14.21 vol. % for raw wood waste and from 12.26 to 15.05 vol. % for torrefied wood waste. The increase in both hydrogen and carbon monoxide gases can be attributed to the endothermic water-gas reaction ($C + H_2O \leftrightarrow CO + H_2$) and the Boudouard reaction ($C + CO_2 \leftrightarrow 2CO$).

Saleh et al. [44] studied the effect of temperature on raw and torrefied empty fruit bunches, and the results show that the synthesis gas yield, lower heating value (LHV), cold gas efficiency (CGE), and tar generation steadily increased for both feedstocks when gasification temperature is increased from 700 to 900°C. Fuchs et al. [56] see the effect of temperature on H_2, CO, CO_2, CH_4, and H_2O content in the product gas. The H_2, H_2O, and CH_4 contents decrease with increasing temperature, whereas the CO and CO_2 content rises with temperature.

Ming et al. [57] studied the steam co-gasification of biosolids with wood pellets and switchgrass using thermogravimetric analysis-mass spectrometry (TGA–MS) and ash as a catalyst. Hydrogen (H_2) production ranged from 48 to 52% by volume at temperatures of 728–825°C, while carbon monoxide (CO) yield was between 10

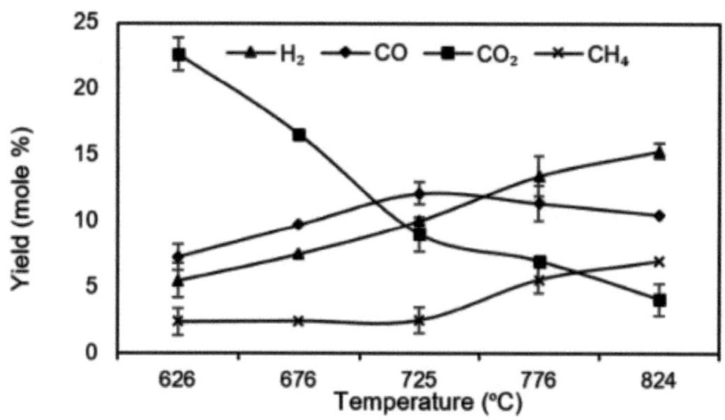

Figure 2.
The effect of temperature on syngas composition (taken from [54]).

and 12%. The increased H_2 content resulted from tar cracking, with tar production at 10.3–200 g/Nm^3. Methane (CH_4) concentration rose from 13.2 to 16.4%, and syngas yield increased from 0.29 to 0.99 m^3/kg as temperature rose. However, higher temperatures were limited by the melting and agglomeration of fuel ash in the gasifier. Tian et al. [58] examined the effects of temperature on syngas yield and composition using an updraft fixed bed reactor. The highest hydrogen (H_2) concentrations from lignin, cellulose, and hemicellulose were 54.0% at 920°C, 43.4% at 1220°C, and 40.1% at 1220°C, respectively. These results suggest that the water-gas shift reaction and methane steam reforming were dominant. **Figure 3a** shows that hydrogen potential (the highest amount of H_2 that can be obtained by WGS and hydrocarbon steam reforming) increased below 1020°C and remained constant above it, indicating a weak temperature dependence for heterogeneous reactions. **Figure 3b** illustrates that the hydrogen yield ratio for cellulose and hemicellulose increased from 920–1120°C, while it remained stable for lignin, highlighting the significance of homogenous reactions.

Figure 3c shows that the hydrogen yield for lignin increased with temperature up to 1020°C due to heterogeneous reactions, with minimal variation above this point. In contrast, cellulose and hemicellulose yields rose from 920 to 1220°C, driven by both reaction types below 1020°C and mainly by homogeneous reactions above it. The highest yields recorded were 0.27 Nm^3/kg for cellulose and 0.30 Nm^3/kg for hemicellulose at 1220°C, and 0.88 Nm^3/kg for lignin at 1020°C.

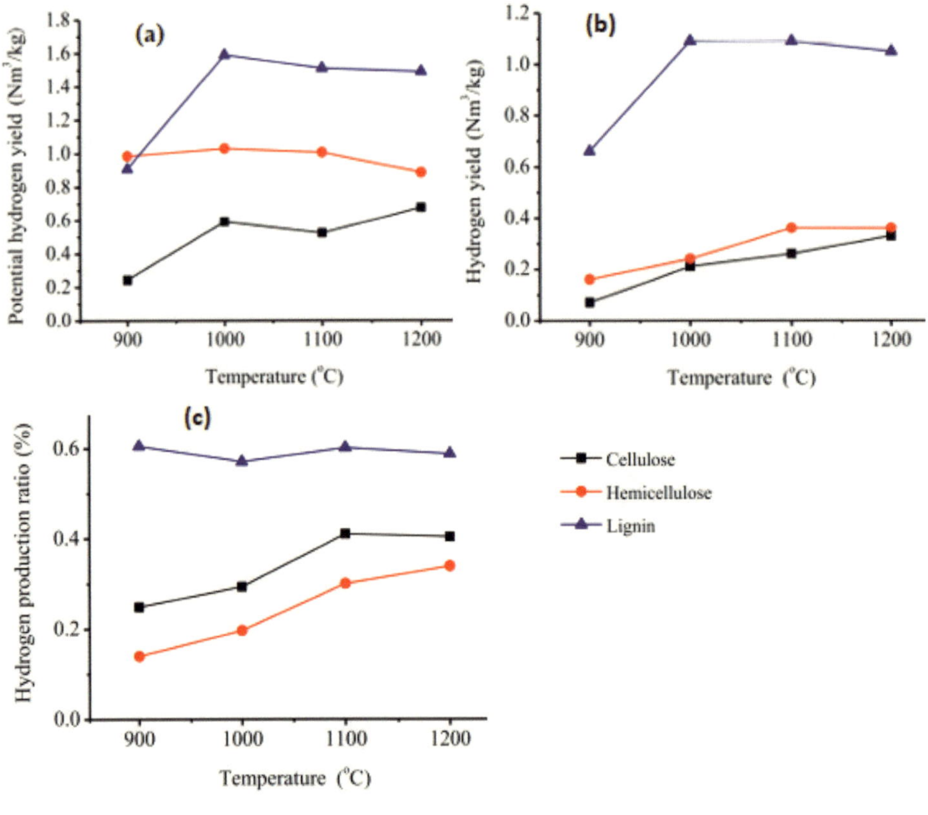

Figure 3.
Hydrogen production parameters of cellulose, hemicellulose, and lignin at different temperatures: (a) potential hydrogen yield; (b) hydrogen production ratio; (c) hydrogen yield (adapted from [58]).

Fan et al. [59] gasified rice husk in a circulating fluidized bed (CFB) biomass gasifier. At lower temperatures, the combustible components of CO, H_2, and CH_4 increase with the increase in temperature, reaching the highest values of 20.35%, 6.85%, and 7.09% at 789°C. Then, with a further increase in temperature, CO, H_2, and CH_4 decrease to 14.33%, 5.23%, and 4.59% at 840°C.

Weiland et al. [60] study the effects of process parameters such as the O_2 stoichiometric ratio (k), the load of the gasifier, the gasifier pressure, and the fuel particle size on the performance of an oxygen-blown entrained flow biomass gasifier. The syngas concentration of CH_4 showed a clear correlation to the process temperature. Huang [61] suggested that the content of CO decreases by 4.17% from 700 to 900°C, and the content of H_2 increases by 5.43%. The fraction of CH_4 decreases by 6.95% as the fraction of CO_2 increases by 5.69%. Lyons et al. [62] examined how gasification temperature affects the main combustible gases (H_2, CO, and CH_4) and CO_2 from spruce, alder, and pine. As the temperature increased from 750 to 950°C, H_2, CO, and CH_4 concentrations rose, while CO_2 concentration fell.

4.3 Effect of equivalence ratio (ER)

The equivalence ratio (ER) is the ratio of actual oxygen supply to fuel, divided by the oxygen needed for complete combustion of theoretical biomass. It affects the biomass feed rate, gasifier supply rate, temperature, pressure, calorific value, and gas composition, depending on the gasifier, gasification agent, and feedstock type [50]. In gasification, the equivalency ratio (ER) affects the blending of the gasifying agent and feedstock, which in turn impacts the stability of the process and the quality of the syngas. This, in turn, influences the feeding rate [63]. As the equivalence ratio increases, the peak explosive pressures initially rise and then subsequently decline [64]. As the equivalence ratio (ER) increases, more oxygen becomes available for the oxidation reaction that generates the necessary energy, thereby raising the gasifier temperature while maintaining a constant biomass flow rate. For gasification, the suitable value of equivalence ratio (ER) ranges from 0.2 to 0.4, allowing for controlled production of tar and char [64]. Finding the optimal ER greatly affects energy, economic, and environmental outcomes [53]. Saleh et al. [44] studied the gasification of raw and torrefied empty fruit bunch (EFB) to assess the impact of temperature and equivalence ratio (ER) on gasification performance and tar generation in an air-fluidized gasifier. As ER increased, synthesis gas yield and cold gas efficiency (CGE) improved for both feedstocks, within the range of ER 0.26 ~ to 0.33. There exists an optimal synthesis gas yield with changes in ER. Up to a certain point, concentrations of H_2 and CO increase, which enhances CGE. Torrefied EFB generated more tar than raw EFB, with yields rising from 16.43 g tar/kg biomass to 20.23 g tar/kg biomass as ER increased from 0.26 to 0.33. Higher ER values supply more air, boosting hydrogen and carbon monoxide production, but also lead to lower CO and H_2 concentrations and higher CO_2 content, reducing syngas heating value. Additionally, gas temperature rises with increasing ER due to exothermic reactions from higher oxygen levels [65].

Jafari et al. [66] assessed an air-blown entrained flow cyclone gasifier and found that increasing the equivalence ratio (ER) from 0.2 to 0.3 (about 1.5 mol/kg of fuel) slightly decreased carbon monoxide (CO) yield and had a negligible effect on carbon dioxide (CO_2) yield. However, hydrogen (H_2) yield increased to around 0.9 mol/kg of fuel, while methane (CH_4) yield decreased by approximately 30%. Huang [61] reported a 6.78% increase in H_2 content with an S/B ratio rise from 0.7 to 2.8, while CO decreased by 19.52%, CH4 by 1.9%, and CO_2 increased by 13.74%. At 850°C, gasification of alder yielded 22.1%, 18.7%, and 12.5% less H_2, CO, and CH_4, respectively, at an equivalence ratio (ER) of 0.19 compared to 0.38. At 950°C, the yields were 10.5%, 5.1%, and 9.7% lower, respectively, at the same ER. H_2 concentration decreased most significantly with a lower ER [62]. Sudarmanta [67] examined how air ratio (AR) and equivalence ratio (ER) affect the performance of a multi-stage downdraft gasifier, finding that higher AR percentages increase gas content and lower the lower heating value (LHV) to 4.221 kJ/kg. The CH_4 composition decreases with increasing equivalence ratio (ER) due to slower hydrogasification reactions at higher temperatures. At an ER of 0.5, CO_2 content rises with temperature in the oxidation zone, benefiting combustion, but CO and H_2 lower heating value (LHV) compositions decrease. This occurs as more combustion air approaches stoichiometric conditions, while gasification is expected to involve combustion reactions. Bambang et al. [68] find that a higher equivalence ratio improves gasification, gas yield, and carbon conversion but decreases the lower heating value of syngas.

4.4 Effect of gasification agent

The gasifying agent is essential in gasification. The choice of gasifying agent depends on the required syngas quality for downstream applications [69]. Gasifying agents yield different reactivities and gas compositions [13]. Hydrogen (H_2) and carbon monoxide (CO) are generated when heavier hydrocarbons interact with gasifying agents, which can include pure oxygen, carbon dioxide, steam, or air, and result in gases with different heating values [50]. The composition and heating value of the product gas depends on the type, quality, and quantity of the gasifying agent used in biomass gasification [51]. The gasifying agent is chosen based on the required syngas quality for downstream applications [69]. Air is often used in gasification processes due to its availability and cost-effectiveness [70]. Islam [71] studied the effects of five gasifying agents—steam, CO_2, H_2O_2, O_2, and air—on syngas compositions at an equivalence ratio (ER) of 0.33. The concentration of carbon monoxide decreased with increasing modified equivalence ratio (MER). CO_2 yielded the highest carbon monoxide concentration, followed by oxygen, H_2O_2, steam, and air. Using CO_2 as a gasifying agent yielded a higher carbon monoxide concentration than oxygen, H_2O_2, steam, and air gasifying agents. H_2O_2 was more effective than the other agents in producing high hydrogen concentrations in syngas.

Mehmood et al. [72] optimize gasifying agents in a 3D downdraft system to improve gas composition, combustion, and CO_2 utilization. They found that a mixture of 15% CO_2 with 23–50% O_2-enriched air yields the highest combustible compounds in syngas. A 50–50% air-steam mixture also promotes CO production. Introducing steam enhances the pyrolysis and gasification of byproducts like tar and coke, improving the overall gasification reaction and product gas quality [61]. Xiang et al. [73] studied the effects of gasifying agents—steam, CO_2, steam + O_2, CO_2 + O_2, steam + CO_2, and steam + CO_2 + O_2—on syngas composition. Increasing the S/B ratio with steam raised H_2 and CO_2 yields but decreased CO and CH_4 yields.

4.5 Effect of particle size

Li et al. [74] discovered that the CO peak rate gradually increases from 0.30 to 0.49, 0.51, and 1.0 mLs^{-1} g^{-1}, respectively, with increasing particle size. This is probably because some small particle samples float on molten salt in the early reaction. Huang [61] found that the particle size changes from 60 mesh to 100 mesh, the proportions of CO and CO2 decrease by 2.25 and 4.84%, respectively, and the proportions of H_2 and CH_4 increase by 6.85 and 0.49%, respectively. The volume ratio of each component gas does not change substantially. Wibowo et al. [75] confirmed that the smaller the particle size, the lower the biomass bed's porosity.

4.6 Moisture content

Alnouss et al. [76] found that in steam gasification, higher moisture content increases hydrogen (H_2) production while reducing carbon monoxide (CO) presence. More steam reacts with carbon to produce H_2, resulting in less conversion to CO_2, CO, and methane (CH_4). Different behavior was demonstrated for CO_2 production for oxygen gasification and oxygen/steam gasification of biomass. The production of H_2, CO, and CH_4 decreases as moisture content increases, leading to a greater conversion of carbon to CO_2. In the case of steam gasification, maintaining a high moisture content is advantageous, as the objective is to maximize H_2 production. Conversely, for oxygen and oxygen/steam gasification, it is preferable to reduce moisture content. To optimize H_2 generation, high moisture is preferred in steam gasification when the gasification temperature can be maintained enough to sustain the reaction. The study also noted a consistent decrease in lower heating value (LHV) with increased moisture content. Ghassemi and Shahsavan-markadeh [77] found that cold gas efficiency decreases with increasing moisture content at lower equivalence ratios, while it remains constant at higher ratios. This is attributed to greater oxygen availability, which enhances exothermic reactions, resulting in complete moisture evaporation and higher gasification temperatures.

4.7 Residence time

Residence time is defined as the average time the molecules spend in the reactor. The gasification efficiency increases to a certain extent when increasing residence time but remains unchanged after a particular value of residence time increment [39]. Hernández et al. [78] studied the gasification of grapevine pruning, sawdust, de-alcoholized grape marc, and a coal-coke blend (50 wt.% each) in an entrained flow gasifier. The highest CO and H2 values were obtained from grapevine pruning wastes, with 23.6% vol. CO and 11.8% vol. H_2 at 1.9 seconds. CH_4 levels were generally stable, except for a slight increase in sawdust. The gas yield (GY) increased slightly for grapevine pruning and de-alcoholized grape marc, with a more notable rise in the sawdust and coal-coke blend. Hernández et al. [79] found that longer residence times in an entrained flow gasifier enhance producer gas quality by increasing the concentrations of combustible species (CO, H_2, and CH_4). This effect is consistent across all biomass fuels, despite the short residence times typical of this gasifier type.

4.8 Biomass feeding rate

It is beneficial to increase the biomass feeding rate (FR) for a specific gasifier to enhance output capacity. However, if the feeding rate is excessively high, it will lead to a shorter gas residence time and increased gas production, which can result in poorer gas quality and a higher tar yield [80]. Factors such as reactor design and biomass properties significantly influence the biomass feeding rate. An excessively high feeding rate, for instance, can lead to clogging, which may reduce conversion efficiency or even interrupt the reaction process. Therefore, while increasing the feeding rate may enhance syngas production, it is essential to determine the optimal rate for the reactor's proper operation to avoid issues such as partial gasification, lower-quality syngas, and degradation of reactor performance [81]. Guo et al. [80] studied the impact of operational parameters on biomass gasification in a downdraft fixed bed system. They found that increasing the biomass feeding rate decreased hydrogen (H_2) and carbon monoxide (CO) concentrations, suggesting that high feeding rates hinder cracking and reforming reactions. Conversely, carbon dioxide (CO_2) levels rose, indicating increased oxidation reactions. The shorter gas residence time in the reduction zone also led to reduced CO_2 consumption through carbon (C) reactions.

5. Challenges and future perspectives

Despite its potential as a clean energy source, bioenergy is underutilized commercially due to challenges in biomass supply chain management and conversion technologies. Biomass, especially agricultural residues, has high moisture content at harvest (over 50%) and requires proper drying for storage, increasing transportation costs. Sourced mainly from rural areas, effective management necessitates community involvement and support through advanced machinery, policies, and marketing. Effective supply chain management is vital for consistent supply to conversion units, and pretreatment methods like drying, densification, and grinding enhance processing efficiency. Solar drying is cost-effective but inefficient under low solar irradiance, while conventional methods provide faster drying rates. A moisture content of ≤40% is optimal for gasification to maximize hydrogen yield. Briquetting is preferred for commercializing biomass due to its advantages in heat transfer, transportation, and storage. However, inadequate infrastructure, information gaps, financial burdens, and policy limitations hinder bioenergy progress. Gasification produces an explosive gaseous mixture, with fuel quality affected by design and operating conditions [9]. Small and medium-scale applications face challenges due to high capital costs, leading to inadequate designs for essential processes like tar removal. While there is demand for eco-friendly technologies, not all biomass gasification methods are financially viable, as seen in straw gasification for domestic cooking [82]. Tar formation significantly challenges biomass gasification, causing equipment blockages and increased maintenance. It is a viscous, dark liquid with a low condensation temperature that can obstruct downstream equipment. Lighter hydrocarbons (C2–6) may form tarry aerosols instead of condensing, adversely affecting gas quality and rendering it unsuitable for high-purity applications beyond boiler use. Effective tar removal and cracking technologies are needed [12].

Moreover, feedstock variability, gas cleanup, and economical viability, including high initial capital costs and operational expenses, can make biomass gasification less competitive compared to other energy sources; technological integration with other processes like combined heat and power (CHP) are also the challenge of biomass gasification. Integration with renewable energy systems, which combine biomass gasification with solar or wind energy, brings hybrid systems that enhance overall energy efficiency and reliability [83]; feedstock pretreatment (such as torrefaction and pelletization) can enhance the uniformity and energy density of biomass feedstocks, improving gasification performance; employing catalysts that can enhance reaction rates and improve the quality of syngas [12]; and application of plasma gasification and microwave-assisted gasification needs future attention for enhancing the overall performance of biomass gasification [42].

6. Conclusion

The adoption of renewable energy sources, such as biomass, solar, and wind, has garnered considerable interest owing to their capacity to reduce environmental impacts and promote sustainable energy generation. Biomass is adaptable and superior to other renewable energy sources like solar, wind, and hydropower since it can be directly transformed into gas, liquid, and solid goods. A possible technique for turning biomass into gaseous fuels is biomass gasification. It is a clean and effective technique as well, and the gas products that are targeted may be utilized as fuels, raw materials for synthesis chemicals, or liquid oil. Comprehending the impact of gasifier types and operation settings on biomass gasification performance is essential. For a better gasification process design and operation, it might also offer helpful information. Several kinds of reactors might be used for biomass gasification. The fluidized bed and the updraft/downdraft moving bed are often used gasifiers because of the relatively high reactivity of biomass and the challenges associated with biomass collection.

Conflict of interest

The authors declare that there is no conflict of interest.

Author details

Tesfa Nega Gesese[1,2]*, Asmare Tezera Admase[1,2], Ejigayehu Desalegn Asrade[1,2] and Eshetu Getahun[1,2]*

1 Bahir Dar Energy Center, Bahir Dar Institute of Technology, Bahir Dar University, Bahir Dar, Ethiopia

2 Chemical Engineering Program, Faculty of Chemical and Food Engineering, Bahir Dar Institute of Technology, Bahir Dar University, Bahir Dar, Ethiopia

*Address all correspondence to: tesfanega510@gmail.com and eshetu201384@gmail.com

IntechOpen

© 2025 The Author(s). Licensee IntechOpen. This chapter is distributed under the terms of the Creative Commons Attribution License (http://creativecommons.org/licenses/by/4.0), which permits unrestricted use, distribution, and reproduction in any medium, provided the original work is properly cited.

References

[1] Sansaniwal SK, Pal K, Rosen MA, Tyagi SK. Recent advances in the development of biomass gasification technology: A comprehensive review. Renewable and Sustainable Energy Reviews. 2017;**72**:363-384. DOI: 10.1016/j.rser.2017.01.038

[2] Dhanak DV. Biomass gasification: A modern approach for renewable energy utilization. GRD Journals- Global Research and Development Journals. 2016;**1**:58-65

[3] Broer KM, Peterson C. Gasification. In: Thermochemical Processing of Biomass: Conversion into Fuels, Chemicals and Power. 2nd ed. John Wiley & Sons Ltd.; 2019

[4] Lourinho G, Garcia B, Rijo B, Brito P. Costs of gasification Technologies for Energy and Fuel Production: Overview, Analysis, and Numerical Estimation. Recycling. 2023;**8**:49

[5] Madadian E, Lefsrud M, Andres C, Lee P, Roy Y. Green energy production: The potential of using biomass gasification. Journal of Green Engineering. 2014;**4**:101-116. DOI: 10.13052/jge1904-4720.421

[6] Ojolo SJ. Techno-economic analysis of biomass energy utilization through gasification technology for sustainable energy production and economic development in Nigeria. Journal of Energy. 2018;**2018**:860252. DOI: 10.1155/2018/4860252

[7] Ebissa DT, Getahun E. Development and performance evaluation of biomass-based Injera baking gasifier stove: A case study of clean cooking Technologies in Ethiopia. Scientific World Journal. 2024;**2024**:1524398

[8] Getahun E, Tessema D, Gabbiye N. Design and development of household gasifier cooking stoves: Natural versus forced draft. In: ICST Institute for Computer Sciences, Social Informatics and Telecommunications Engineering. Switzerland AG: Springer International Publishing; 2019. pp. 298-314

[9] Molino A, Chianese S, Musmarra D. Biomass gasification technology: The state of the art overview. Journal of Energy Chemistry. 2016;**25**:10-25. DOI: 10.1016/j.jechem.2015.11.005

[10] Safarian S, Unnþórsson R, Richter C. A review of biomass gasification modelling. Renewable and Sustainable Energy Reviews. 2019;**110**:378-391. DOI: 10.1016/j.rser.2019.05.003

[11] Hosseinpour J, Chitsaz A, Liu L, Gao Y. Simulation of eco-friendly and affordable energy production via solid oxide fuel cell integrated with biomass gasification plant using various gasification agents. Renewable Energy. 2020;**145**:757-771. DOI: 10.1016/j.renene.2019.06.033

[12] Sikarwar VS, Zhao M, Clough P, Yao J, Zhong X, Memon MZ, et al. An overview of advances in biomass gasification. Energy & Environmental Science. 2016;**9**:2939-2977. DOI: 10.1039/c6ee00935b

[13] Andre Y, Zhao Z, Yoshida A, Abudula A, Guan G. Small-scale biomass gasification systems for power generation (< 200 kW class): A review. Renewable and Sustainable Energy Reviews. 2020;**117**:109486. DOI: 10.1016/j.rser.2019.109486

[14] Sandoval LP, Díaz CA. Sustainability aspects of biomass gasification systems

for small power generation. Renewable and Sustainable Energy Reviews. 2020;**134**:110180. DOI: 10.1016/j.rser.2020.110180

[15] Kuyumani EM, Mambo S, Taziwa RT. Biomass Gasification for Sustainable Renewable Electricity Generation. 2024

[16] Asadullah M. Barriers of commercial power generation using biomass gasification gas: A review. Renewable and Sustainable Energy Reviews. 2014;**110180**(29):201-215. DOI: 10.1016/j.rser.2013.08.074

[17] Gesese TN, Getahun E, Getahun AA. Investigation of thermal degradation properties and chemical kinetic characteristics of biomass pyrolysis via TG / DTG and FTIR techniques: Sesame stalks as potential source of bioenergy in Ethiopia. International Journal of Energy Research. 2024;**2024**:8891126. DOI: 10.1155/2024/8891126

[18] Bridgwater AV. The technical and economic feasibility of biomass gasification for power generation. Fuel. 1995;**74**:631-653

[19] Buragohain B, Mahanta P, Moholkar VS. Biomass gasification for decentralized power generation: The Indian perspective. Renewable and Sustainable Energy Reviews. 2010;**14**:73-92. DOI: 10.1016/j.rser.2009.07.034

[20] Richardson Y, Drobek M, Julbe A, Blin J, Pinta F. Biomass gasification to produce syngas. In: Recent Advances in Thermo-Chemical Conversion of Biomass. Elsevier; 2015. pp. 213-250

[21] Basu P. Biomass Gasification and Pyrolysis. Elsevier; 2010

[22] Giglio E, Vitale G, Lanzini A, Santarelli M. Biomass and bioenergy integration between biomass gasification and high-temperature electrolysis for synthetic methane production. Biomass and Bioenergy. 2021;**148**:106017. DOI: 10.1016/j.biombioe.2021.106017

[23] Phounglamcheik A, Vila R, Kienzl N, Wang L, Hedayati A, Brostr M, et al. CO_2 Gasification Reactivity of Char from High-Ash Biomass. Washington, DC: American Chemical Society; 2021. DOI: 10.1021/acsomega.1c05728

[24] Freda C, Catizzone E, Villone A, Cornacchia G. Biomass gasification in rotary kiln integrated with a producer gas thermal cleaning unit: An experimental investigation. Results in Engineering. 2024;**21**:101763. DOI: 10.1016/j.rineng.2024.101763

[25] Ahrenfeldt J, Thomsen TP, Henriksen U, Clausen LR. Biomass gasification cogeneration e a review of state of the art technology and near future perspectives. Applied Thermal Engineering. 2013;**50**:1407-1417. DOI: 10.1016/j.applthermaleng.2011.12.040

[26] Ebissa DT, Getahun E. Evaluation of biomass-based cook stoves in Ethiopia: A transition strategy from conventional to modern cook stoves. Environmental Progress & Sustainable Energy. 2024;**44**:e14531

[27] Seitarides T, Athanasiou C, Zabaniotou AÃ. Modular biomass gasification-based solid oxide fuel cells (SOFC) for sustainable development. Renewable and Sustainable Energy Reviews. 2008;**12**:1251-1276. DOI: 10.1016/j.rser.2007.01.020

[28] Siriwardhana M, Salam PA, Kumar S. The status of biomass gasification in Thailand and Cambodia: Prepared for Energy Environment Partnership on in Thailand and Cambodia Prepared for: Energy Environment Partnership (EEP),

Mekong Region. 2010. DOI: 10.13140/RG.2.2.16424.67844

[29] Benti NE, Gurmesa GS, Argaw T, Aneseyee AB, Gunta S, Kassahun GB, et al. The current status, challenges and prospects of using biomass energy in Ethiopia. Biotechnology for Biofuels. 2021;**14**:1-24. DOI: 10.1186/s13068-021-02060-3

[30] Ding L, Yang M, Dong K, Viet D, Douglas NV, Jiahan H, et al. Mobile power generation system based on biomass gasification. International Journal of Coal Science & Technology. 2022;**9**:34. DOI: 10.1007/s40789-022-00505-0

[31] Rahman M, Henriksen UB, Ciolkosz D. Startup process, safety and risk assessment of biomass gasification for off - grid rural electrification. Scientific Reports. 2023;**13**:1-9. DOI: 10.1038/s41598-023-46801-w

[32] Mai TP, Nguyen DQ. Gasification of biomass. In: Biotechnological Applications of Biomass. London, UK: IntechOpen; 2020

[33] Zhang K, Chang J, Guan Y, Chen H, Yang Y, Jiang J. Lignocellulosic biomass gasification technology in China. Renewable Energy. 2013;**49**:175-184. DOI: 10.1016/j.renene.2012.01.037

[34] Sadhwani N, Adhikari S, Eden MR. Biomass gasification using carbon dioxide: Effect of temperature, CO_2/C ratio, and the study of reactions influencing the process. Industrial and Engineering Chemistry Research. 2016;**55**:2883-2891. DOI: 10.1021/acs.iecr.5b04000

[35] Sikarwar VS. Biomass Gasification. Elsevier; 2017

[36] Mandal S, Daguppati S, Bandyopadhyay R, Das AK. Recent advances in biomass gasification: A review. In: Macromolecular Characterization of Hydrocarbons for Sustainable Future, Green Energy and Technology. Singapore: Springer Nature; 2021

[37] Loha C, Karmakar MK, De S. Gasifiers: Types, operational principles, and commercial forms. In: Coal and Biomass Gasification, Energy, Environment, and Sustainability. Singapore: Springer Nature; 2018. pp. 63-91

[38] Bukar AA, Ben OM, Tela BM, Eljummah AM. Assessment of biomass gasification: A review of basic design considerations. American Journal of Energy Research. 2019;**7**:1-14. DOI: 10.12691/ajer-7-1-1

[39] Mishra S, Upadhyay RK. Review on biomass gasification: Gasifiers, gasifying mediums, and operational parameters. Materials Science for Energy Technologies. 2021;**4**:329-340. DOI: 10.1016/j.mset.2021.08.009

[40] Hegazi AA, Okasha FM. A comprehensive review of biomass gasification process. International Journal of Scientific and Engineering Research. 2019;**10**:266-299

[41] Agustina TE, Faizal M. The effect of air fuel ratio and temperature on syngas composition and calorific value produced from downdraft gasifier of rubber wood-coal mixture. International Journal of Engineering. 2018;**31**:1480-1486

[42] Kaushal R, Amit R, Dhaka K. A comprehensive review of the application of plasma gasification technology in circumventing the medical waste in a post-COVID-19 scenario. Biomass Conversion and Biorefinery. 2024;**14**:1427-1442. DOI: 10.1007/s13399-022-02434-z

[43] Shahbaz M, Al-ansari T, Inayat M, Sulaiman SA, Parthasarathy P, Mckay G. A critical review on the influence of process parameters in catalytic co-gasification: Current performance and challenges for a future prospectus. Renewable and Sustainable Energy Reviews. 2020;**134**:110382. DOI: 10.1016/j.rser.2020.110382

[44] Saleh S, Asma N, Abdul F. Effects of gasification temperature and equivalence ratio on gasification performance and tar generation of air fluidized bed gasification using raw and torrefied empty fruit bunch. Chemical Engineering Transactions. 2021;**88**:1309-1314. DOI: 10.3303/CET2188218

[45] Yaghoubi E, Xiong Q, Doranehgard MH, Yeganeh MM, Shahriari G, Bidabadi M. The effect of different operational parameters on hydrogen rich syngas production from biomass gasification in a dual fluidized bed gasifier. Chemical Engineering and Processing Process Intensification. 2018;**126**:210-221. DOI: 10.1016/j.cep.2018.03.005

[46] Vamvuka D, Zacheila K. Gasification performance of barley straw waste blended with lignite for syngas production under steam or carbon dioxide atmosphere. Applied Sciences. 2024;**14**:665

[47] Ozonoh M, Oboirien BO, Higginson A, Daramola MO. Performance evaluation of gasification system efficiency using artificial neural network. Renewable Energy. 2020;**145**:2253e2270. DOI: 10.1016/j.renene.2019.07.136

[48] Faizal FZMCKM, Atnaw NAFASSM. Gasification performance of sawdust, pelletized sawdust and sub - bituminous coal in a downdraft gasifier. SN Applied Sciences. 2020;**2**:1-10. DOI: 10.1007/s42452-020-03358-x

[49] Saad J, Williams PT. Manipulating the H_2/CO ratio from dry reforming of simulated mixed waste plastics by the addition of steam. Fuel Processing Technology. 2017;**156**:331-338. DOI: 10.1016/j.fuproc.2016.09.016

[50] Gao Y, Wang M, Raheem A, Wang F, Wei J, Xu D, et al. Syngas production from biomass gasification: Influences of feedstock properties, reactor type, and reaction parameters. ACS Omega. 2023;**8**:31620-31631. DOI: 10.1021/acsomega.3c03050

[51] Zain M, Mahshab S, Toor UA. Recent approaches on the optimization of biomass gasification process parameters for product H_2 and syngas ratio: A review. Environment, Development and Sustainability. 2022:1-29. DOI: 10.1007/s10668-022-02279-6

[52] Inayat A, Ahmad MM, Mutalib MIA. Effect of process parameters on hydrogen production and efficiency in biomass steam gasification with in-situ CO_2 capture. In: Proceedings of International Conference on Process Engineering and Advanced Materials. 2010

[53] Cao Y, Bai Y, Du J. Air-steam gasification of biomass based on a multi-composition multi-step kinetic model: A clean strategy for hydrogen-enriched syngas production. Science of the Total Environment (Netherlands: Elsevier). 2021;**753**:141690. DOI: 10.1016/j.scitotenv.2020.141690

[54] Khezri R, Ghani WAWAK, Biak ADR, Yunus R, Silas K. Experimental evaluation of Napier grass gasification in an autothermal bubbling fluidized bed reactor. Energies. 2019;**12**:1517

[55] Jamin NA, Saleh S, Asma N, Abdul F. Influences of gasification temperature and equivalence ratio on fluidized

bed gasification of raw and torrefied wood wastes. Chemical Engineering Transactions. 2020;**80**:127-132. DOI: 10.3303/CET2080022

[56] Fuchs J, Schmid JC, Müller S, Mauerhofer AM, Benedikt F, Hofbauer H. The impact of gasification temperature on the process characteristics of sorption enhanced reforming of biomass. Biomass Conversion and Biorefinery. 2020;**10**:925-936

[57] Ming M, Masnadi MS, Grace JR, Bi XT, Lim CJ, Li Y. Co-gasification of biosolids with biomass: Thermogravimetric analysis and pilot scale study in a bubbling fluidized bed reactor. Bioresource Technology. 2015;**175**:51-58. DOI: 10.1016/j.biortech.2014.10.045

[58] Tian T, Li Q, He R, Tan Z. Effects of biochemical composition on hydrogen production by biomass gasification. International Journal of Hydrogen Energy. 2017;**42**:1-10. DOI: 10.1016/j.ijhydene.2017.06.174

[59] Fan X, Yang L, Jiang J. Experimental study on industrial-scale CFB biomass gasification. Renewable Energy. 2020;**158**:32-36. DOI: 10.1016/j.renene.2020.05.145

[60] Weiland F, Wiinikka H, Hedman H, Wennebro J, Pettersson E. Influence of process parameters on the performance of an oxygen blown entrained flow biomass gasifier. Fuel. 2015;**153**:510-519. DOI: 10.1016/j.fuel.2015.03.041

[61] Huang F, SJ. Investigation of biomass (pine wood) gasification: Experiments and Aspen plus simulation. Energy Science & Engineering. 2019;7:1-10. DOI: 10.1002/ese3.338

[62] Lyons A, Konist A, Lees H, Järvik O. Effect of woody biomass gasification process conditions on the composition of the producer gas. Sustainability. 2021;**13**:11763

[63] Sidek FN, Samad NAFA, Saleh S. Review on effects of gasifying agents, temperature and equivalence ratio in biomass gasification process review on effects of gasifying agents, temperature and equivalence ratio in biomass gasification process. IOP Conference Series: Materials Science and Engineering. 2020;**863**:012028. DOI: 10.1088/1757-899X/863/1/012028

[64] Bisht AS, Thakur NS. Small scale biomass gasification plants for electricity generation in India: Resources, installation, technical aspects, sustainability criteria & policy. Renewable Energy Focus. 2019;**28**:112-126. DOI: 10.1016/j.ref.2018.12.004

[65] Ismail TM, El-salam MA. Parametric studies on biomass gasification process on updraft gasifier high temperature air gasification. Applied Thermal Engineering. 2017;**112**:1460-1473. DOI: 10.1016/j.applthermaleng.2016.10.026

[66] Jafari PH, Wingren A, Hellström JGI, Rikard B. Effect of process parameters on the performance of an air-blown entrained flow cyclone gasifier. International Journal of Sustainable Energy ISSN. 2020;**39**:21-40. DOI: 10.1080/14786451.2019.1626858

[67] Sudarmanta ARS, Sudarmanta B. Experimental investigation on multi-stage downdraft gasification: Influence of air ratio and equivalent ratio to the gasifier performance. In: AIP Conference Proceedings. AIP Publishing; 2018. pp. 020026-1-020026-6

[68] Sudarmanta B, Gafur A, Saleh AR, Bambang Arip Dwiyantoro S. The

effect of two stage gasifying agent on biomass downdraft gasification to the gasifier performance. AIP Conference Proceedings. 2018;**1983**:020037-1-020037-6

[69] Mansur FZ, Faizal CKM. Review on effects of gasifying agents, temperature and equivalence ratio in biomass gasification process review on effects of gasifying agents, temperature and equivalence ratio in biomass gasification process. IOP Conference Series: Materials Science and Engineering. 2020;**863**:012028. DOI: 10.1088/1757-899X/863/1/012028

[70] Susastriawan AAP, Saptoadi H. Small-scale downdraft gasifiers for biomass gasification: A review. Renewable and Sustainable Energy Reviews. 2017;**76**:989-1003. DOI: 10.1016/j.rser.2017.03.112

[71] Islam W. Effect of different gasifying agents (steam, H_2O_2, oxygen, CO_2, and air) on gasification parameters. International Journal of Hydrogen Energy. 2020;**45**:31760-31774. DOI: 10.1016/j.ijhydene.2020.09.002

[72] Mehmood A, Tahir MW, Saeed MA, Arshad MY, Hussain H, Mularski J, et al. Optimization of gasifying agents in 3D downdraft gasification for enhanced gas composition, combustion, and CO_2 utilization. Fire. 2023;**6**:361

[73] Xiang Y, Cai L, Guan Y, Liu W, Cheng Z, Liu Z. Study on the effect of gasification agents on the integrated system of biomass gasification combined cycle and oxy-fuel combustion. Energy. 2020;**206**:118131. DOI: 10.1016/j.energy.2020.118131

[74] Li J, Xie Y, Zeng K, Flamant G, Yang H, Yang X, et al. Biomass gasification in molten salt for syngas production. Energy. 2020;**210**:118563. DOI: 10.1016/j.energy.2020.118563

[75] Wibowo WA, Pranolo SH, Sunarno JN, Purwadi D. Effect of biomass feed size and air flow rate on the pressure drop of Jurnal Teknologi full paper effect of biomass feed size and air flow rate on the pressure drop of gasification reactor. Jurnal Teknologi (Sciences & Engineering). 2014;**68**:7-12. DOI: 10.11113/jt.v68.2962

[76] Alnouss A, Mckay G, Al-ansari T. Production of syngas via gasification using optimum blends of biomass. Journal of Cleaner Production. 2020;**242**:118499. DOI: 10.1016/j.jclepro.2019.118499

[77] Ghassemi H, Shahsavan-markadeh R. Effects of various operational parameters on biomass gasification process; a modified equilibrium model. Energy Conversion and Management. 2014;**79**:18-24. DOI: 10.1016/j.enconman.2013.12.007

[78] Hernández JJ, Ballesteros R, Aranda G. Characterisation of tars from biomass gasification: Effect of the operating conditions. Energy. 2013;**50**:333-342. DOI: 10.1016/j.energy.2012.12.005

[79] Hernández JJ, Aranda-almansa G, Bula A. Gasification of biomass wastes in an entrained flow gasifier: Effect of the particle size and the residence time. Fuel Processing Technology. 2010;**91**:681-692. DOI: 10.1016/j.fuproc.2010.01.018

[80] Guo F, Dong Y, Dong L, Guo C. Effect of design and operating parameters on the gasification process of biomass in a downdraft fixed bed: An experimental study. International Journal of Hydrogen Energy. 2014;**39**:5625-5633. DOI: 10.1016/j.ijhydene.2014.01.130

[81] Inayat A, Rocha-meneses L, Jamil F, Ghenai C, Shanableh A. Paramaeters

That Affect Biomass Gasification. Encyclopedia. pp. 4-9. Available from: https://encyclopedia.pub/entry/42801

[82] Leung DYC, Yin XL, Wu CZ. A review on the development and commercialization of biomass gasification technologies in China. Renewable and Sustainable Energy Reviews. 2004;**8**:565-580. DOI: 10.1016/j.rser.2003.12.010

[83] Walter V, Lisa G, Ahlstr JM. The role of biomass gasification in the future flexible power system—BECCS or CCU? Renewable Energy. 2022;**190**:596-605. DOI: 10.1016/j.renene.2022.03.100

Chapter 5

Catalytic Gasification of Biomass: Materials, Products and Other Considerations

Sergio O. Flores Valle, Daniel Martín Márquez López, Fredy Josealdo Castillo Plata, Ektaí López Ángeles and Gerardo de los Santos Camas

Abstract

This chapter describes briefly and consciously the state of art of the different catalytic materials, their characterization, catalytic behavior and reaction conditions applied in experimental studies of biomass gasification process in concordance with the kind of product obtained. It also considers modern criteria like sustainability, scalability, economic feasibility, process optimization and trends in catalysts' development with potential commercial applications.

Keywords: gasification, biomass, catalytic materials, syngas, hydrogen

1. Introduction

Biomass and products are those considered in the production of syngas, hydrogen (H_2), methane (CH_4), pyrolytic oils and value-added products. Gasification process is an efficient and clean path that converts different biomasses into mostly valuable products, such as syngas, hydrogen, methane and pyrolytic oils. Although the produced gas can be used as a fuel in downstream equipment, power machines and electrochemical fuel cells, it is also possible to obtain valuable chemical products like dimethyl ether and methanol. In general, conversion efficiency of biomass depends on multiple factors, such as biomass characteristics and conversion methods. In biological processes, such as anaerobic digestion, biogas and residuals digested are produced by the activity of microbial organism. In contrast, thermochemical methods apply heat and pressure to transform biomass into biofuels, gases and chemicals. Thermochemical processes are among the most applied techniques in biomass conversion for biofuel production [1]. This chapter excludes biochemical processes of gasification and analyzes the production of methane, hydrogen, syngas and pyrolytic oils by biomass gasification technology based on its technical characteristics, which is the main obstacle to the implementation of gasification technologies at high scale.

 IntechOpen

2. Process description

Pyrolysis as the first step of the gasification process produces release of volatile species and char formation. It is followed by the char reactions with a gasifying agent to obtain hydrogen, methane and carbon dioxide (CO_2) and carbon monoxide (CO). It is known that steam gasification has attracted more interest, as a higher hydrogen yield can be reached [2]. Several biomass gasification studies have focused on the production of direct combustion gases based on gasification parameters and gasifier performance. It is known that methane, the main component of natural gas, is one of the primary sources of energy extensively employed worldwide [3].

3. Types of gasifiers

Gasifiers can be categorized into three main types: moving or fixed bed, fluidized bed and entrained flow. Updraft fixed bed gasifiers produce syngas characterized by a high tar content, while downdraft fixed bed gasifiers, though beneficial for biomass conversion, are limited to small-scale operations. The primary obstacle to using the generated gas as fuel in downstream processes is the tar, which is an unwanted byproduct. Entrained flow gasifiers require a finely powdered feedstock, making them unsuitable for biomass gasification without extensive pretreatment methods such as torrefaction or pyrolysis. Fluidized bed gasifiers are advantageous for biomass conversion due to their flexibility in fuel types and scalability. A specific type within this category is the steam-blown dual fluidized bed (DFB) gasifier, which produces high-quality syngas that is both high in heating value and nitrogen-free, with various designs currently being developed [4].

4. Methane production by gasification

According to the experiments carried out by Wang [5] using an indirectly heated, fluidized bed biomass gasifier, it was observed that longer residence times lead to a product gas composition that more closely approaches equilibrium conditions. Specifically, hydrogen (H_2) and carbon monoxide (CO) concentrations rise, while methane (CH_4) and other light hydrocarbons decrease as the temperature rises. Also, they determined that methane and other light hydrocarbons remain virtually the same over all equivalent ratios (ERs) (from 0.2 to 0.4) tested. Their research demonstrated that methane (CH_4) production declines as temperature increases under an equivalent ratio (ER) of 0.3. At the same time, carbon conversion efficiency improves significantly with higher temperatures, a result of faster chemical reaction rates. Furthermore, gas yield rises substantially with increasing temperature, primarily due to reduced char formation and the transformation of complex hydrocarbons into simpler diatomic molecules, such as carbon monoxide (CO) and hydrogen (H_2). Moreover, within the range of steam-to-biomass ratios tested, they found that hydrogen concentration increases, carbon monoxide levels decrease slightly and carbon dioxide (CO_2) remains relatively stable as the amount of steam rises. Conversely, the production of methane and other light hydrocarbons experiences a slight reduction as the steam-to-biomass ratio increases. In summary, the production of methane is favored out of equilibrium. However, longer residence time and higher temperatures improve the extent of gasification reactions.

These experimental data showed agreement with the simulation of gasification of animal wastes done by employing Aspen Plus to determine the selection of the best operating conditions from an environmental point of view [6]. It demonstrated that when using steam and CO_2 as gasifying agents, the production of H_2 and CO was enhanced at high temperatures, while the formation of CH_4 and CO_2 was favored at lower temperature. Another notable effect was that the increase in the gasifying agent/biomass ratio had a negative influence on the production of methane (CH_4).

However, biomass char gasification to high CH_4 concentration faces challenges in current technology. It is important to remark that another alternative to promote the production of methane over other gasification products is the use of the appropriate catalyst, to promote methanation reaction, which plays a key role in increasing methane production. Typically, compounds of transition metals like iron (Fe), cobalt (Co), nickel (Ni) and cerium (Ce) are noted for their strong catalytic performance in gasification reactions [7]. Ni-based catalyst is also used to efficiently catalyze methanation at 250–500°C [8]. Additionally, catalysts used for the hydrogenation of carbon dioxide to methane demonstrated high CO_2 conversion rates along with selectivity for CH_4. The benefits of nickel metal include its availability and low cost, as well as its effective performance [9].

$$CO + 3H_2 \rightarrow H_2O + CH_4 \quad (1)$$

$$CO_2 + 4H_2 \rightarrow 2H_2O + CH_4 \quad (2)$$

Zhaoa [10] reported the production of methane-free excess glycerol through coupling of steam reforming and methanation carried out in a fixed-bed system with two catalytic reactors at atmospheric conditions, using nickel-manganese/aluminum oxide (Ni-Mn/Al_2O_3) catalyst, enhanced by the addition of 3 wt% Mn exhibiting an excellent catalytic stability and a high activity, resulting in an 87% CH_4 yield during glycerol steam reforming (GSR) syngas methanation.

In 2022, Si et al. [11] proposed a catalytic-driven method for the direct conversion of solid biomass into bio-natural gas using a catalyst containing the nickel-aluminum (Ni2Al3) alloy phase. This approach achieved nearly complete conversion of various agricultural and forestry residues, with the total carbon yield of gas products reaching up to 93% after several hours at a relatively low temperature of 300°C. We can describe this process as economically competitive based on a low-carbon footprint estimated by a preliminary life cycle assessment, especially for non-fossil hydrogen and low hydrogen pressure.

Likewise, it is well established that high methane (CH_4) selectivity can be achieved at pressures greater than 1 atm and relatively low temperatures. This is due to the fact that CO methanation is both a volume-reducing and exothermic reaction. To prevent carbon deposition during the initial screening of methanation catalysts, the reaction temperature should be maintained at a low level. However, such operating conditions are generally not recommended for industrial applications due to practical and efficiency considerations [12].

This argument is funded by the research of Minowa et al. [13], who directly gasified a water slurry of cellulose into methane using reduced nickel on kieselguhr and sodium carbonate as catalysts at a temperature of 400°C and a pressure of approximately 13 MPa for 1 hour, conducted in a conventional 100-mL stainless-steel autoclave.

Gasifier	Temperature °C	Pressure (atm)	Catalyst	CH_4 yield %
Bed reactors [12]	200	1	Nickel-based catalysts	100
Fixed bed [10]	275	1	Ni/Al_2O_3 $Ni\text{-}Mn_x/Al_2O_3$	87.5
Two-stage fixed-bed reactor [14]	750	1	Ni-Co/PWA	56
Fixed bed [7]	750	34.54	K-modified transition metal composite catalysts	90

Table 1.
Comparison of conditions and performances of gasifiers.

In addition, Zhang et al. [14] utilized a metal-doped activated char from pine wood (PWA), specifically Ni-Co/PWA, in the catalytic reforming of pyrolysis vapor from pine sawdust in a hydrogen atmosphere to enhance tar removal and increase the production of methane-rich gas. It obtained a tar yield as low as 1.2 wt.% and methane of 56.0%. Moreover, in the same year, Jiao et al. [7] studied the methane production from high-pressure catalytic steam hydrogasification of sawdust Char on K-modified transition metal composite catalysts. Gasification experiments were carried out in a self-designed high-pressure fixed-bed reactor at high temperature (700, 750 and 800°C) and pressure (0.1, 1.5, 2.5 and 3.5 MPa). Noticing that with pressure from 0.1 to 3.5 MPa, it was possible to increase CH_4 yield by 9.14 times, because pressurization converts CO_2 and CO into CH_4 to a certain extent [7]. Conditions and performances of gasifiers of some relevant studies described above are summarized in **Table 1**.

5. Production of hydrogen energy from biomass gasification

Hydrogen assists as an alternative for future energy strategies, potentially replacing natural gas (methane) as an energy source due to its distinctive characteristics as a clean and non-toxic fuel with a high energy yield of 122 kJ per kg, which is 2.75 times higher than those of other hydrocarbon fuels. Hydrogen is considered a highly efficient alternative to gasoline, as just 9.5 kg of hydrogen can effectively replace 25 kg of gasoline, showing the potential to decarbonize industries, such as transportation, iron and steel production [15].

Hydrogen can be produced from different processes, such as: steam reforming, partial oxidation of hydrocarbons, pyrolysis, fermentation, biophotolysis, wind power, hydropower, electrolysis and gasification [16].

Hydrogen production from biomass gasification as well-established thermochemical process that does not involve burning could contribute to reduce environmental impact. This method heats biomass, steam and air or oxygen to temperatures exceeding 700°C in a controlled environment to obtain H_2, CO, CO_2 and CH_4. Subsequent to gasification stage, a water-gas shift reaction takes place, producing additional hydrogen and carbon dioxide. The extraction of hydrogen from the gas mixture can be accomplished through the use of adsorbers or specialized membranes. As well, biomass provides renewable and sustainable energy in other forms, such as syngas, biogas and producer gas. Gasification also produces solid residues (char and/or soot) and other undesired products such as tar (heavy hydrocarbons) [17].

The selection of the appropriate gasification agent is made based on a balance between the final product gas composition and process costs. Biomass gasification can be categorized into indirect (allothermal) and direct (autothermal) methods, based on how heat is supplied to the system. In air- or oxygen-blown gasifiers (autothermal), the necessary heat for the reaction is generated internally through exothermic oxidation reactions that occur directly within the gasifier. Applying air as an oxidizing agent leads to obtain a lower heating value (LHV) syngas containing 7–12% of H_2 percentage, air is largely used in pilot plants due to commercial and economical interest [16, 18]. The use of pure oxygen (O_2) combined with steam as a gasification agent has been explored to reduce the nitrogen (N_2) concentration in syngas. This approach results in a high-quality gas mixture composed of approximately 40% hydrogen (H_2), 40% carbon monoxide (CO) and 20% carbon dioxide (CO_2). Additionally, the process achieves minimal tar and char content, attributed to the higher temperatures reached during gasification. However, the operating costs are high due to those costs involved by O_2 production [15]. Particular efforts have been done in order to improve H_2 production by paying special attention to the variables involved in the process, as demonstrated by Chen et al. who found the optimized H_2 yield of 41.36 mol % efficiency at 757.65°C with an equivalent ratio (ER) of 0.241 for 22.36 min to produce a high-quality syngas at a moderate temperature and air as a gasifying agent [18].

On the other hand, allothermal gasifiers usually use steam as a gasification agent. An external heat source is needed to maintain the required operating temperature typical range from 670–900°C and is determined by chemical reaction constraints regarding carbon conversion, tar reforming, possible inherent capture of sulfur species and CO_2, as well as by agglomeration and fouling through ash melting depending on the fuel composition. The use of pure steam produces syngas with high H_2 concentration around 40%, 25% CO, 25% CO_2 and 8% CH_4, with the amount of tar generated being moderate [19, 20].

The ongoing advancement in gasification technology has led to the development of various gasifier designs, including fixed bed (such as updraft and downdraft), fluidized bed (including bubbling, circulating, dual and multi-stage configurations), entrained flow, spouted bed and plasma reactors. These designs cater to different operational needs and improve the efficiency and versatility of gasification processes [21]. According to Jürgen and Pröll, steam gasification in dual fluidized bed (DFB) gasifiers is the most promising technology to produce hydrogen and syngas from solid fuels in medium and small-scale applications [20].

Factors like pretreatments, biomass conversion, gasifier performance, limited hydrogen yield, purification and separation technologies represent the main obstacle to enhance the efficiency of hydrogen generation from biomass [17]. However, tar formation during the gasification process remains the most significant challenge hindering hydrogen production and the commercialization of biomass gasification technology. This is because syngas with high tar content can lead to pipeline fouling, downstream corrosion, catalyst deactivation and negative impacts on health and the environment. To address this issue, catalytic materials are utilized to improve hydrogen production and effectively eliminate tar [16, 19].

Transition metal-based catalysts have shown promising results, especially Ni-based catalysts, as showed by Zhang et al. who employed the ICI46-1, Z409 and RZ409 commercial catalysts with nickel oxide (NiO) as the active component to eliminate heavy tars (>99% destruction efficiency) and increasing hydrogen concentration by 6–11 vol% [22]. Additionally, Ali Abedi demonstrated that the 4Ce/10Ni Al_2O_3 catalyst achieved the highest syngas yield of 2.6 mol/kg daf (dissolved air flotation) pellets and the lowest tar yield of 2.1%. This improvement was attributed to the incorporation of cerium (Ce) as a promoter in the nickel (Ni)-based catalyst, which enhanced metal dispersion and lowered the reduction temperature. These factors contributed to increased syngas production and reduced tar formation [23].

6. Syngas production

Synthesis gas (syngas) represents a promising alternative energy source to conventional fossil fuels, enabling electricity production and finding applications in oil refining plants, the chemical industry and hydrogen production. Hydrogen, in particular, has garnered significant attention as a sustainable fuel and energy carrier, with its market demand anticipated to grow substantially in the coming years. This growing demand can be addressed through a variety of technological approaches, and biomass gasification is emerging as one such method capable of converting organic materials into synthesis gas in an environmentally sustainable manner [24].

The syngas production process can be divided into several key stages. In the feedstock preparation phase, heating, purification and addition of additives such as steam or CO_2 are carried out to bring the feed material to optimum condition. Pyrolysis involves thermal decomposition of the material at high temperatures (200 to 600°C), generating volatile products and a solid residue called "Char." In the partial oxidation stage, the volatile products and part of the char react with limited oxygen, producing CO, H_2O and CO_2, and releasing heat for subsequent reactions. Reduction or gasification consists of reactions where the remaining char reacts with water vapor and CO_2, generating H_2 and CO, and may include secondary reactions, such as water-gas transformation (WGSR) and methanation. Subsequently, purification removes impurities such as hydrogen sulfide (H_2S) and CO_2 from the syngas prior to its industrial use, thus ensuring its suitability for various applications. Finally, in the chemical synthesis stage, syngas is used to produce chemical products and fuels, such as methanol and synthetic fuels, depending on the H_2/CO ratio and the operating conditions required for the final product. These stages make it possible to transform various types of feedstocks into syngas, a valuable intermediate for numerous industrial and energy applications [25].

Gasifiers are devices used to produce syngas by gasifying various fuels, such as coal, biomass or waste. Types of gasifiers include fixed-bed gasifiers, which include the updraft gasifier, where the fuel is fed from above and the gasifier from below, generating a syngas with high CO_2 concentration; the downdraft gasifier, where the gasifier enters from the top and the gas exits from the bottom, producing a better-quality syngas; and the cross-current gasifier, which combines features of both. Fluidized bed gasifiers include the bubbling fluidized bed gasifier, which allows good mixing and heat transfer, and the circulating fluidized bed gasifier, which provides more precise control and high capacity. There are also entrained flow gasifiers, which operate at high temperatures and produce syngas with low tar content. Other types include plasma gasifiers, which use plasma energy at high temperatures and are effective but expensive, and rotary gasifiers, which facilitate the movement of solids but can generate more tar. In addition, there are gasifiers designed specifically for biomass or waste, which require special technologies to clean the syngas due to the higher number of contaminants. The choice of the right gasifier depends on the type of fuel and the specific operating conditions [26].

The system of chemical reactions that generate syngas includes several key reactions during gasification and reforming, resulting in a mixture composed mainly of hydrogen (H_2) and carbon monoxide (CO). The main reactions involved in syngas production are described below in four groups:

1. Carbon gasification reactions where organic or fossil materials are converted to syngas by high temperature reactions with a controlled amount of oxygen and/or steam, are especially important for biomass feedstocks. We find Partial Oxidation (3), which consists of reacting hydrocarbons with a limited amount of oxygen to directly produce synthesis gas, stands out for the speed of the process and the efficiency in producing synthesis gas from heavier hydrocarbons [27]; Gasification with steam (4) and Boudouard Reaction (5).

$$C + 0.5O_2 \rightarrow CO\ \Delta H = -111\,kJ/mol \tag{3}$$

$$C + H_2O \rightarrow CO + H_2\ \Delta H = +131\,kJ/mol \tag{4}$$

$$C + CO_2 \leftrightarrow 2CO\ \Delta H = +172\,kJ/mol \tag{5}$$

2. Reforming Reactions consist of reacting hydrocarbons with steam to produce synthesis gas [28]; Steam Methane Reforming (6); Dry Methane Reforming (7); and Partial Methane Oxidation (8).

$$CH_4 + H_2O \rightarrow CO + 3H_2\ \Delta H = +206\,kJ/mol \tag{6}$$

$$CH_4 + CO_2 \leftrightarrow 2CO + 2H_2\ \Delta H = +247\,kJ/mol \tag{7}$$

$$CH_4 + 0.5O_2 \rightarrow CO + 2H_2\ \Delta H = -36.6\,kJ/mol \tag{8}$$

3. Water-Gas Shift Reaction (9) [29].

$$CO + H_2O \leftrightarrow CO_2 + H_2\ \Delta H = -41.2\,kJ/mol \tag{9}$$

4. Methanation Reactions (10) and (11) [29].

$$2CO + 2H_2 \leftrightarrow CH_4 + CO_2\ \Delta H = -247\,kJ/mol \tag{10}$$

$$CO + 3H_2 \leftrightarrow CH_4 + H_2O\ \Delta H = -206\,kJ/mol \tag{11}$$

Syngas production is a complex thermodynamic process involving high temperatures and pressures, as well as the interaction of various operating conditions that affect the structure and equilibration of the system, thus optimizing the ratio of H_2 to CO required for specific applications, such as methanol synthesis or synthetic fuel production.

The gasification of syngas is a complex process that requires several operating conditions and parameters that affect the quality of the gas produced. Operating temperatures in gasifiers typically range from 450 to 1400°C; higher temperatures

Category	Type	Description	Advantages and Disadvantages
Noble Metals	Platinum (Pt) [31]	Used in various oxidation and reforming reactions, it offers high activity and selectivity.	Advantage: High activity; Disadvantage: High cost.
Noble Metals	Ruthenium (Ru) [32]	Used in hydrogen production, favorable under high pressure and temperature conditions.	Advantage: High efficiency; Disadvantage: High price.
Transition Metals	Nickel (Ni) [33]	Common in methane reforming and gasification, provides good stability and activity.	Advantage: Low cost; Disadvantage: Sensitive to deactivation.
Metal Oxides	Al_2O_3 [34]	Used as a support in catalysts, it improves the dispersion of active metals.	Advantage: Good support; Disadvantage: Little activity on its own.
Metal Oxides	ZnO [35]	Acts as a catalyst to produce light hydrocarbons, improves selectivity.	Advantage: Effectiveness at low temperatures; Disadvantage: Poor thermal stability.
Zeolites	ZSM-5 [36]	Acid catalyst, used in pyrolysis and production of aromatic compounds.	Advantage: High selectivity; Disadvantage: Prone to clogging.
Zeolites	HY MCM-41 [37]	Used in adsorption and as a support for metals in catalytic processes, improves the surface.	Advantage: High activity; Disadvantage: High production costs.
Carbon Based (Allotrope)	Activated carbon [38]	Used in adsorption and as a support for metals in catalytic processes, improves the surface.	Advantage: Low cost; Disadvantage: Low durability.
Perovskites	$LaCoO_3$ [39]	Used in the oxidation and reduction of organic compounds, noted for its versatility.	Advantage: Versatility; Disadvantage: High production costs.
Carbide	Co_2C [40]	Used for various catalytic processes.	Advantage: High activity; Disadvantage: Cost variability based on source.
Transition Metal Oxides	Vanadium oxide supported catalysts [41]	Used in oxidation processes, mainly for the synthesis of organic compounds.	Advantage: Cost-effective; Disadvantage: Limited application scope.

Table 2.
Employed catalysts in hydrogen production by biomass gasification.

favor carbon conversion. Pressure, which generally varies between 1 and 70 bar, also has a significant impact, improving carbon conversion and reducing tars, although it increases equipment complexity. The equivalence ratio (ER), crucial for establishing substoichiometric conditions, ensures efficient gasification. The gasifying agents, which can be air, oxygen or steam, have a decisive influence on the syngas composition: air produces a gas with high N_2 content and low calorific value, while oxygen generates a syngas richer in H_2 and CO, and steam favors hydrogen formation. The syngas composition depends on the type of fuel used and the operating conditions, as well as the carbon conversion efficiency, the production of tars and other performance parameters such as production rate and gas quality. Finally, it is crucial to implement cleaning technologies to remove contaminants such as H_2S and CO_2, ensuring the suitability of the syngas for energy and chemical applications. Optimization of these parameters is essential to maximize syngas production and improve its viability [30].

Catalytic processes are critical for improving syngas conversion efficiency. Several innovations have been developed as shown in **Table 2**, such as nickel (Ni) catalysts, which are active in CO_2-CH_4 reforming reactions, although they face carbon deposition deactivation challenges. Bimetallic double layer oxides (LDO) and cobalt carbides (Co_2C) present opportunities in methane reforming and hydrocarbon formation. Also, strategies such as the double bed allow efficient conversion of syngas to methanol and then to light kerosene. In addition, vanadium oxide-supported catalysts have demonstrated high activity in selective methanol oxidation, while catalytic pyrolysis of low-density plastics with Zeolite Socony Mobil-5 (ZSM-5), HY and MCM-41 catalysts shows the potential of these materials for applications in syngas conversion. These advances are key to offering sustainable solutions in the production of high-value chemical compounds using renewable resources and managing carbon emissions, thus improving the sustainability of syngas-related industrial processes [42].

7. Pyrolytic oil production

The pyrolysis of biomass has emerged as a promising thermochemical conversion process for producing renewable liquid fuels, particularly bio-oil, over the past decade. Pyrolysis involves the thermal decomposition of biomass in the absence of oxygen, yielding three primary products: bio-oil (a liquid fraction), biochar (a solid residue) and non-condensable gases. Among these, bio-oil has garnered significant attention due to its potential as a substitute for fossil-derived fuels and as a feedstock for chemical production. Recent advancements in understanding the thermodynamics, kinetics and economic aspects of biomass pyrolysis have provided valuable insights into optimizing the process for commercial viability. Excellent revisions are found in the literature [43–47].

8. Thermodynamics of biomass pyrolysis

The thermodynamics of biomass pyrolysis focuses on energy changes and equilibrium conditions during thermal decomposition. Recent studies have highlighted the importance of temperature, pressure and biomass composition in determining the enthalpy and entropy changes associated with pyrolysis. The process typically requires temperatures between 400 and 600°C to initiate the breakdown of biomass components, such as cellulose, hemicellulose and lignin [48–50]. Thermodynamic models have been developed to predict the optimal conditions for maximizing bio-oil yield, with particular emphasis on minimizing energy losses and maximizing process efficiency. For instance, the presence of moisture in biomass has been shown to significantly affect the energy balance, as the evaporation of water consumes additional thermal energy, thereby reducing overall efficiency [51]. Furthermore, the use of advanced thermodynamic simulations has enabled researchers to identify conditions that favor the formation of bio-oil over other byproducts, such as biochar and gases [52].

9. Kinetics of biomass pyrolysis

The kinetics of biomass pyrolysis, which describes the rate of thermal decomposition, has been extensively studied to improve process control and product distribution. Recent research has focused on developing detailed kinetic models that account for the complex reactions involved in the breakdown of biomass components. These models have revealed that temperature is the most critical factor influencing reaction rates, with higher temperatures not only accelerating decomposition but also promoting the formation of non-condensable gases at the expense of bio-oil [53]. Additionally, the use of catalysts has emerged as a promising strategy to modify reaction pathways and enhance the quality of bio-oil. Acidic and basic catalysts, for example, have been shown to reduce the oxygen content of bio-oil, improving its stability and energy density. These advancements in kinetic modeling and catalytic pyrolysis have significantly contributed to the optimization of bio-oil production.

10. Economic viability of biomass pyrolysis

The economic feasibility of biomass pyrolysis depends on several factors, including feedstock costs, process efficiency and the market value of the products. Recent economic analyses have estimated the production cost of bio-oil to range between $0.5 and $1.5 per liter, depending on the scale of operation and geographic location [54]. The market value of bio-oil is closely tied to its quality, with higher-quality oils commanding premium prices as fuel or chemical feedstocks. Advances in posttreatment techniques, such as hydrodeoxygenation, have improved the quality of bio-oil, making it more competitive with conventional fuels [55]. Moreover, the valorization of byproducts, such as biochar and non-condensable gases, has enhanced the overall economics of the process. Biochar can be sold as a soil amendment or carbon sequestration material, while the gases can be used to generate heat or electricity, offsetting operational costs [56, 57]. Despite these advancements, achieving economies of scale remains a critical challenge for the widespread adoption of biomass pyrolysis.

Over the past decade, significant progress has been made in improving the yield and quality of bio-oil from biomass pyrolysis. One of the most notable advancements is the development of catalytic pyrolysis, which employs catalysts to tailor the composition of bio-oil, making it more suitable for use as a fuel or chemical feedstock. Additionally, integrated process schemes that combine pyrolysis with other thermochemical conversion technologies, such as gasification, have demonstrated potential for enhancing overall efficiency and reducing costs [58]. Pilot-scale and demonstration plants have been established in various regions, providing valuable data on the technical and economic feasibility of scaling up biomass pyrolysis [59]. These efforts have underscored the importance of process optimization and the need for sustainable biomass sourcing to ensure long-term viability [60].

11. Challenges and future perspectives

Despite the progress made, several challenges remain to be addressed for the commercialization of biomass pyrolysis. Process optimization, particularly in terms of energy efficiency and product yield, requires further research. Additionally, ensuring the sustainability of biomass feedstocks and minimizing the environmental impact of pyrolysis are critical considerations. Future research should focus on developing cost-effective catalysts, improving process integration and scaling up production to achieve economies of scale [61, 62]. Furthermore, policy support and market incentives will play a crucial role in promoting the adoption of bio-oil as a renewable fuel and chemical feedstock [63, 64].

Author details

Sergio O. Flores Valle*, Daniel Martín Márquez López, Fredy Josealdo Castillo Plata, Ektaí López Ángeles and Gerardo de los Santos Camas
Instituto Politécnico Nacional, México City, México

*Address all correspondence to: sfloresv@ipn.mx

IntechOpen

© 2025 The Author(s). Licensee IntechOpen. This chapter is distributed under the terms of the Creative Commons Attribution License (http://creativecommons.org/licenses/by/4.0), which permits unrestricted use, distribution, and reproduction in any medium, provided the original work is properly cited.

References

[1] Gu H, Tang Y, Yao J, Chen F. Study on biomass gasification under various operating conditions. Journal of the Energy Institute. 2019;**92**(5):1329-1336. DOI: 10.1016/j.joei.2018.10.002

[2] Sharma S, Sheth PN. Air–steam biomass gasification: Experiments, modeling and simulation. Energy Conversion and Management. 2016;**110**:307-318. DOI: 10.1016/j.enconman.2015.12.030

[3] Wu Y, Ye X, Wang Y, Wang L. Methane production from biomass by thermochemical conversion: A review. Catalysts. 2023;**13**(4):771-786. DOI: 10.3390/catal13040771

[4] Doherty W, Reynolds A, Kennedy D. Aspen plus simulation of biomass gasification in a steam blown dual fluidised bed. In: Materials and Processes for Energy: Communicating Current Research and Technological Developments. Dublin: Formatex Research Centre; 2013. pp. 212-220

[5] Wang Y, Kinoshita CM. Experimental analysis of biomass gasification with steam and oxygen. Solar Energy. 1992;**49**(3):153-158. DOI: 10.1016/0038-092X(92)90066-J

[6] Fernandez-Lopez M, Pedroche J, Valverde JL, Sanchez-Silva L. Simulation of the gasification of animal wastes in a dual gasifier using Aspen plus®. Energy Conversion and Management. 2017;**140**:211-217. DOI: 10.1016/j.enconman.2017.03.008

[7] Jiao W, Ding X, Jiao W, Liu Y, Wang Z, Fang Y. Methane production from high-pressure catalytic steam hydrogasification of sawdust char on K-modified transition metal composite catalysts. Fuel. 2024;**363**:131000. DOI: 10.1016/j.fuel.2024.131000

[8] Zhang J, Wang M, Xu S, Feng Y. Hydrogen and methane mixture from biomass gasification coupled with catalytic tar reforming, methanation and adsorption enhanced reforming. Fuel Processing Technology. 2019;**192**:147-153. DOI: 10.1016/j.fuproc.2019.04.023

[9] Tamimi K, Alavi SM, Rezaei M, Akbari E. Preparation of the Mn-promoted NiO–Al_2O_3 nanocatalysts for low temperature CO_2 methanation. Journal of the Energy Institute. 2021;**99**:48-58. DOI: 10.1016/j.joei.2021.08.008

[10] Zhao H, Xu C, Wang T. Production of methane from biomass glycerol through coupling of steam reforming and methanation on Ni-Mn/Al_2O_3. Sustainable Chemistry and Pharmacy. 2019;**13**:100150. DOI: 10.1016/j.scp.2019.100150

[11] Si X, Lu R, Zhao Z, Yang X, Wang F, Jiang H, et al. Catalytic production of low-carbon footprint sustainable natural gas. Nature Communications. 2022;**13**(1):258. DOI: 10.1038/s41467-021-27919-9

[12] Gao J, Wang Y, Ping Y, Hu D, Xu G, Gu F, et al. A thermodynamic analysis of methanation reactions of carbon oxides for the production of synthetic natural gas. RSC Advances. 2012;**2**(6):2358-2368. DOI: 10.1039/C2RA00632D

[13] Minowa T, Ogi T, Dote Y, Yokoyama S. Methane production from cellulose by catalytic gasification. Climate Change Energy and the Environment. 1994;**5**(5):813-815. DOI: 10.1016/0960-1481(94)90094-9

[14] Zhang M, Hu B, Fan G, Yang M, Lu Q, Wu Y. The removal of tar and the production of methane-rich gas from biomass hydrogen pyrolysis by using biochar-based catalysts. Energy Conversion and Management. 2024;**313**:118596. DOI: 10.1016/j.enconman.2024.118596

[15] Parthasarathy P, Narayanan KS. Hydrogen production from steam gasification of biomass: Influence of process parameters on hydrogen yield – A review. Renewable Energy. 2014;**66**:570-579. DOI: 10.1016/j.renene.2013.12.025

[16] Alptekin FM, Celiktas MS. Review on catalytic biomass gasification for hydrogen production as a sustainable energy form and social, technological, economic, environmental, and political analysis of catalysts. ACS Omega. 2022;**7**(29):24918-24941. DOI: 10.1021/acsomega.2c01538

[17] Obiora NK, Ujah CO, Asadu CO, Kolawole FO, Ekwueme BN. Production of hydrogen energy from biomass: Prospects and challenges. Green Technologies and Sustainability. 2024;**2**(3):100100. DOI: 10.1016/j.grets.2024.100100

[18] Chen G, Jamro IA, Samo SR, Wenga T, Baloch HA, Yan B, et al. Hydrogen-rich syngas production from municipal solid waste gasification through the application of central composite design: An optimization study. International Journal of Hydrogen Energy. 2020;**45**(58):33260-33273. DOI: 10.1016/j.ijhydene.2020.09.118

[19] Cortazar M, Santamaria L, Lopez G, Alvarez J, Zhang L, Wang R, et al. A comprehensive review of primary strategies for tar removal in biomass gasification. Energy Conversion and Management. 2023;**276**:116496. DOI: 10.1016/j.enconman.2022.116496

[20] Karl J, Pröll T. Steam gasification of biomass in dual fluidized bed gasifiers: A review. Renewable and Sustainable Energy Reviews. 2018;**98**:64-78. DOI: 10.1016/j.rser.2018.09.010

[21] Sikarwar VS, Zhao M. Biomass gasification. In: Abraham MA, editor. Encyclopedia of Sustainable Technologies. Elsevier; 2017. pp. 205-216. DOI: 10.1016/B978-0-12-409548-9.10533-0

[22] Zhang R, Brown RC, Suby A, Cummer K. Catalytic destruction of tar in biomass derived producer gas. Energy Conversion and Management. 2004;**45**(7):995-1014. DOI: 10.1016/j.enconman.2003.08.016

[23] Abedi A, Dalai AK. Steam gasification of oat hull pellets over Ni-based catalysts: Syngas yield and tar reduction. Fuel. 2019;**254**:115585. DOI: 10.1016/j.fuel.2019.05.168

[24] Mahinpey N, Abdalla A, Farooqui A. Chapter 9—Biomass gasification for hydrogen production: A pathway to cleaner energy transition. In: Nanda S, Dalai AK, editors. Biomass to Bioenergy. Woodhead Publishing; 2024. pp. 205-235. DOI: 10.1016/B978-0-443-15377-8.00009-6

[25] Flórez-Orrego D, Domingos M, Nakashima R, Teles M, De Oliveira Junior S. Chapter 3. Ammonia production from syngas. In: Section I: Chemicals Production from Syngas. Vol. 3. Amsterdam: Elsevier; 2022. p. 490

[26] Rahimpour MR, Makarem MA, Meshksar M, editors. Index. In: Advances in Synthesis Gas: Methods, Technologies and Applications. Vol. 2. Elsevier; 2023. pp. 449-452. DOI: 10.1016/B978-0-323-91877-0.09990-6

[27] Zhang Y, Zhang S, Gossage JL, Lou HH, Benson TJ. Thermodynamic

analyses of tri-reforming reactions to produce syngas. Energy & Fuels. 2014;**28**(4):2717-2726. DOI: 10.1021/ef500084m

[28] Luneau M, Gianotti E, Meunier FC, Mirodatos C, Puzenat E, Schuurman Y, et al. Deactivation mechanism of Ni supported on Mg-Al spinel during autothermal reforming of model biogas. Applied Catalysis B: Environmental. 2017;**203**:289-299. DOI: 10.1016/j.apcatb.2016.10.023

[29] Farniaei M, Rahnama H, Abbasi M, Rahimpour MR. Simultaneous production of two types of synthesis gas by steam and tri-reforming of methane using an integrated thermally coupled reactor: Mathematical modeling. International Journal of Energy Research. 2014;**38**(10):1260-1277. DOI: 10.1002/er.3125

[30] Rahimpour MR, Makarem MA, Meshksar M, editors. Index. In: Advances in Synthesis Gas: Methods, Technologies and Applications. Vol. 3. Elsevier; 2023. pp. 465-468. DOI: 10.1016/B978-0-323-91878-7.09990-1

[31] Alawi NM, Al-Mohammedawi HH, Nguyen HM, Azeez RA, Shams OA, Sukkar KA. Catalysts for reforming of methane (a review). Petroleum Chemistry. 2024;**64**(8):964-971. DOI: 10.1134/S0965544124070089

[32] Ahluwalia AS. Chapter 3—Synthetic fuels derived from carbon dioxide rich syngas produced through the gasification of biomass. In: Upadhyayula S, Chaudhary A, editors. Sustainable and Green Catalytic Processes for Renewable Fuel Production with Net-Zero Emissions. Elsevier; 2025. pp. 65-81. DOI: 10.1016/B978-0-443-21899-6.00003-3

[33] Wu L, Xie X, Ren H, Gao X. A short review on nickel-based catalysts in dry reforming of methane: Influences of oxygen defects on anti-coking property. In: International Conference of Chemical Engineering & Industrial Biotechnology. Vol. 42. Amsterdam: Elsevier; 2021. pp. 153-160. DOI: 10.1016/j.matpr.2020.10.697

[34] Suttikul T, Nuchdang S, Rattanaphra D, Phalakornkule C. Influence of operating parameters, Al_2O_3 and Ni/Al_2O_3 catalysts on plasma-assisted CO_2 reforming of CH_4 in a parallel plate dielectric barrier discharge for high H_2/CO ratio syngas production. Plasma Chemistry and Plasma Processing. 2020;**40**(6):1445-1463. DOI: 10.1007/s11090-020-10118-7

[35] Kang J, He S, Zhou W, Shen Z, Li Y, Chen M, et al. Single-pass transformation of syngas into ethanol with high selectivity by triple tandem catalysis. Nature Communications. 2020;**11**(1):827. DOI: 10.1038/s41467-020-14672-8

[36] Soares Dias AP, Rego F, Fonseca F, Casquilho M, Rosa F, Rodrigues A. Catalyzed pyrolysis of SRC poplar biomass. Alkaline carbonates and zeolites catalysts. Energy. 2019;**183**:1114-1122. DOI: 10.1016/j.energy.2019.07.009

[37] Chong CC, Bukhari SN, Cheng YW, Setiabudi HD, Jalil AA, Phalakornkule C. Robust Ni/dendritic fibrous SBA-15 (Ni/DFSBA-15) for methane dry reforming: Effect of Ni loadings. Applied Catalysis A: General. 2019;**584**:117174. DOI: 10.1016/j.apcata.2019.117174

[38] Khiari B, Jeguirim M, Limousy L, Bennici S. Biomass derived chars for energy applications. Renewable and Sustainable Energy Reviews. 2019;**108**:253-273. DOI: 10.1016/j.rser.2019.03.057

[39] Lee M, Lim HS, Kim Y, Lee JW. Enhancement of highly-concentrated

hydrogen productivity in chemical looping steam methane reforming using Fe-substituted $LaCoO_3$. Energy Conversion and Management. 2020;**207**:112507. DOI: 10.1016/j.enconman.2020.112507

[40] Arnold RA, Hill JM. Catalysts for gasification: A review. Sustainable Energy & Fuels. 2019;**3**(3):656-672. DOI: 10.1039/C8SE00614H

[41] Wachs IE. Catalysis science of supported vanadium oxide catalysts. Dalton Transactions. 2013;**42**(33):11762-11769. DOI: 10.1039/C3DT50692D

[42] Meng F, Mo W. Recent trends in catalysis for syngas production and conversion. Catalysts. 2023;**13**(9):1284-1286. DOI: 10.3390/catal13091284

[43] Dada TK, Sheehan M, Murugavelh S, Antunes E. A review on catalytic pyrolysis for high-quality bio-oil production from biomass. Biomass Conversion and Biorefinery. 2023;**13**(4):2595-2614. DOI: 10.1007/s13399-021-01391-3

[44] El Bari H, Fanezoune CK, Dorneanu B, Arellano-Garcia H, Majozi T, Elhenawy Y, et al. Catalytic fast pyrolysis of lignocellulosic biomass: Recent advances and comprehensive overview. Journal of Analytical and Applied Pyrolysis. 2024;**178**:106390. DOI: 10.1016/j.jaap.2024.106390

[45] Cai J, Lin N, Li Y, Xue J, Li F, Wei L, et al. Research on the application of catalytic materials in biomass pyrolysis. Journal of Analytical and Applied Pyrolysis. 2024;**177**:106321. DOI: 10.1016/j.jaap.2023.106321

[46] Gollakota ARK, Shu C-M, Sarangi PK, Shadangi KP, Rakshit S, Kennedy JF, et al. Catalytic hydrodeoxygenation of bio-oil and model compounds—Choice of catalysts, and mechanisms. Renewable and Sustainable Energy Reviews. 2023;**187**:113700. DOI: 10.1016/j.rser.2023.113700

[47] Verma P, editor. Thermochemical and Catalytic Conversion Technologies for Future Biorefineries. 1st ed. Springer; 2022. DOI: 10.1007/978-981-19-4312-6

[48] Awad MI, Makkawi Y, Hassan NM. Yield and energy Modeling for biochar and bio-oil using pyrolysis temperature and biomass constituents. ACS Omega. 2024;**9**(16):18654-18667. DOI: 10.1021/acsomega.4c01646

[49] Afrah BD, Riady MI, Arsadha JP, Rimadhina R, Cundari L, ‘Izzah RZ. Effect of pyrolysis temperature and biomass composition on bio-oil characteristics. Ecological Engineering & Environmental Technology. 2024;**25**(3):264-274. DOI: 10.12912/27197050/181152

[50] Fan X, Li B, Yu S, Kang M, Wu H, Bian J, et al. Understanding Pyrolysis of Biomass Waste Residues: Pyrolysis Behaviour, Kinetics, Thermodynamics, and Mechanism Investigations. China: Elsevier; 2023. DOI: 10.2139/ssrn.4593440

[51] Mahmood H, Moniruzzaman M, Yusup S, Khan MI, Khan MJ. Kinetic modeling and optimization of biomass pyrolysis for bio-oil production. Energy Sources, Part A: Recovery, Utilization, and Environmental Effects. 2016;**38**(14):2065-2071. DOI: 10.1080/15567036.2015.1007404

[52] Reza MS, Iskakova ZB, Afroze S, Kuterbekov K, Kabyshev A, Bekmyrza KZ, et al. Influence of catalyst on the yield and quality of bio-oil for the catalytic pyrolysis of biomass: A comprehensive review. Energies. 2023;**16**(14):5547-5585. DOI: 10.3390/en16145547

[53] Soka O, Oyekola O. A feasibility assessment of the production of char using the slow pyrolysis process. Heliyon. 2020;**6**(7):e04346. DOI: 10.1016/j.heliyon.2020.e04346

[54] Machado H, Cristino AF, Orišková S, Galhano dos Santos R. Bio-oil: The next-generation source of chemicals. Reactions. 2022;**3**(1):118-137. DOI: 10.3390/reactions3010009

[55] Afshar M, Mofatteh S. Biochar for a sustainable future: Environmentally friendly production and diverse applications. Results in Engineering. 2024;**23**:102433. DOI: 10.1016/j.rineng.2024.102433

[56] Campbell RM, Anderson NM, Daugaard DE, Naughton HT. Financial viability of biofuel and biochar production from forest biomass in the face of market price volatility and uncertainty. Applied Energy. 2018;**230**:330-343. DOI: 10.1016/j.apenergy.2018.08.085

[57] Mohammed IY, Abakr YA, Mokaya R. Integrated biomass thermochemical conversion for clean energy production: Process design and economic analysis. Journal of Environmental Chemical Engineering. 2019;**7**(3):103093. DOI: 10.1016/j.jece.2019.103093

[58] Yek PNY, Wan Mahari WA, Kong SH, Foong SY, Peng W, Ting H, et al. Pilot-scale co-processing of lignocellulosic biomass, algae, shellfish waste via thermochemical approach: Recent progress and future directions. Bioresource Technology. 2022;**347**:126687. DOI: 10.1016/j.biortech.2022.126687

[59] Trinh QT, Banerjee A, Ansari KB, Dao DQ, Drif A, Binh NT, et al. Upgrading of bio-oil from biomass pyrolysis: Current status and future development. In: Nanda S, Vo D-VN, Sarangi PK, editors. Biorefinery of Alternative Resources: Targeting Green Fuels and Platform Chemicals. Singapore: Springer; 2020. pp. 317-353. DOI: 10.1007/978-981-15-1804-1_14

[60] Sharma A, Pareek V, Zhang D. Biomass pyrolysis—A review of modelling, process parameters and catalytic studies. Renewable and Sustainable Energy Reviews. 2015;**50**:1081-1096. DOI: 10.1016/j.rser.2015.04.193

[61] Carrasco Diaz A, Abdelouahed L, Brodu N, Montes-Jiménez V, Taouk B. Upgrading of pyrolysis bio-oil by catalytic hydrodeoxygenation, a review focused on catalysts, model molecules, deactivation and reaction routes. in Preprints. 2024. DOI: 10.20944/preprints202407.1495.v1

[62] Lachos-Perez D, Martins-Vieira JC, Missau J, Anshu K, Siakpebru OK, Thengane SK, et al. Review on biomass pyrolysis with a focus on bio-oil upgrading techniques. Analytica. 2023;**4**(2):182-205. DOI: 10.3390/analytica4020015

[63] Makepa DC, Chihobo CH, Ruziwa WR, Musademba D. A systematic review of the techno-economic assessment and biomass supply chain uncertainties of biofuels production from fast pyrolysis of lignocellulosic biomass. Fuel Communications. 2023;**14**:100086. DOI: 10.1016/j.jfueco.2023.100086

[64] Osman AI, Farghali M, Ihara I, Elgarahy AM, Ayyad A, Mehta N, et al. Materials, fuels, upgrading, economy, and life cycle assessment of the pyrolysis of algal and lignocellulosic biomass: A review. Environmental Chemistry Letters. 2023;**21**(3):1419-1476. DOI: 10.1007/s10311-023-01573-7

Chapter 6

Fluidization and Fast Pyrolysis Analyses of a Top-Fed Bubbling Fluidized Bed Reactor Using Biomass

Ali Can Sivri

Abstract

The pyrolysis process in a bubbling fluidized bed reactor (BFBR) involves complex, high-temperature multiphase flow phenomena that are not yet fully understood, particularly when using binary mixtures of biomass and bed materials. Accurately predicting fluidization hydrodynamics is critical for optimizing reaction efficiency. This study investigates the influence of the key fluidization parameters on the hydrodynamic behavior and pyrolysis outcomes in a deep BFBR equipped with a novel top-feeding system. Nitrogen was employed both as a transport gas for delivering biomass feedstock and as the fluidizing gas for high-temperature pyrolysis experiments. The results revealed a significant enhancement in the H_2/CO ratio during fast pyrolysis. While the pyrolysis reaction in a fixed bed yielded an H_2/CO ratio of 0.35, the same feedstock processed in the BFBR achieved a nearly 2.5 times increase, underscoring substantial improvements in pyrolysis efficiency.

Keywords: biomass, pyrolysis, bubbling fluidized bed, hydrodynamics, gasification, syngas

1. Introduction

The global energy landscape is undergoing a significant transformation, driven by the urgent need to address rising energy demands, depleting fossil fuel reserves, and the environmental consequences of greenhouse gas emissions. According to the International Energy Agency, energy consumption worldwide is projected to increase by over 20% in the next two decades, underscoring the necessity for sustainable and renewable energy alternatives. Biomass, an abundant and renewable resource, has emerged as a key player in this transition, offering potential for sustainable energy production while mitigating climate change impacts.

Biomass conversion technologies, especially thermochemical methods like pyrolysis and gasification, have gained widespread attention for their ability to transform lignocellulosic materials into valuable energy products. Pyrolysis, a thermal decomposition process conducted in the absence of oxygen, typically operating at temperatures between 300°C and 600°C, is particularly promising for producing bio-oil, biochar, and syngas. The product distribution is influenced by factors such as temperature, heating rate, and feedstock composition. Temperature, in particular, plays a crucial role in the gas yield and its composition. Higher pyrolysis temperatures tend to increase the yield of syngas, with hydrogen content generally rising at temperatures closer to 600°C and above. At lower temperatures, the gas yield is typically lower, and the hydrogen composition is less pronounced, as more carbon is retained in the biochar phase [1, 2].

Fast pyrolysis, which involves rapid heating and short residence times, is recognized as an efficient process for maximizing bio-oil yields. Sulaiman et al. [3] reported that under optimal conditions, including temperatures near 450°C, bio-oil yields could exceed 44%. Zanzi [4] highlighted that operating at higher temperatures (800–1000°C) favors syngas production while minimizing char formation. Pyrolysis occurs in two distinct stages: an initial rapid phase where volatile compounds are released, followed by a slower phase characterized by the rearrangement of residual char [2]. Secondary reactions, such as tar cracking and gas-phase reforming, are strongly influenced by reactor design and operating parameters [5].

Waheed et al. [6] further explored the fast and slow pyrolysis of materials like wood, rice husk, and forestry residue in a fixed-bed reactor, examining how factors such as heating rate, temperature, and steam addition affect product yields and composition. The study showed that fast pyrolysis resulted in significantly higher gas yields, with correspondingly lower oil and char production compared to slow pyrolysis. Increasing the fast pyrolysis temperature (750–1050°C) led to higher gas yields, achieving up to approximately 90% conversion, while oil and char yields declined. At elevated temperatures, the hydrogen concentration in the gas increased, whereas the concentrations of CH_4, CO_2, and $C_2 - C_4$ hydrocarbons decreased.

Among various reactor configurations, bubbling fluidized bed reactors (BFBRs) stand out for their efficiency and versatility in biomass conversion. Unlike fixed-bed reactors, BFBRs fluidize solid particles to create a dynamic environment that enhances heat and mass transfer. The performance of BFBRs is influenced by multiple factors, including particle properties, bed materials, and operating parameters. Particle characteristics, such as size and density, influence bubble formation and distribution, which in turn determine the mixing quality. Spherical particles generally generate smaller and more uniform bubbles, whereas non-spherical particles create larger and more irregular bubbles [7]. Larger and denser particles require higher fluidizing gas rates and tend to mix more slowly than smaller and lighter ones [8]. Biomass particles, with their irregular shape, wide size distribution, and low density, pose challenges in mixing. As a result, bed materials such as sand or glass beads are often used to enhance mixing and heat transfer between particles. A study by Si and Guo [9] demonstrated that adding sand improved the fluidization behavior of biomass particles by increasing particle density and reducing voidage and that the particle size ratio and sand proportion significantly affected fluidization. Particle size distribution (PSD) also significantly impacts fluidization regimes. Wide PSDs tend to transition from bubbling or slugging to turbulent fluidization more easily than narrow PSDs, especially at higher gas velocities [10]. Narrow PSDs show less variation in conversion across different regimes, except in fast fluidization [10].

Bed aspect ratio is another important parameter that influences reaction efficiency. Higher aspect ratios improve reaction efficiency by increasing gas residence time and interparticle attraction but may lead to poorer mixing due to the transition from bubbling to slug-flow regimes [11]. The bed aspect ratio has a negligible effect on the minimum fluidization velocity, as shown by Formisani et al. [12], who found no effect on the minimum fluidization velocity when the aspect ratio was changed from 0.7 to 2.4 in a biomass binary mixture. Homogeneous mixing is essential for optimizing particle interactions, heat transfer, and syngas composition, ensuring uniform distribution of reactants and temperature within the reactor. Uneven mixing can lead to localized hot spots or inefficient syngas production, thus undermining overall reaction efficiency. However, providing quasi-homogeneous mixing in top-fed deep-bed applications remains a challenge. Wilk et al. [13] investigated the impact of feeding position on gasification efficiency and found that top feeding (on-bed feeding) resulted in reduced gasification efficiency compared to bottom and side feeding due to less effective mixing of the inert material and feedstock particles. Despite their effectiveness, conventional feeding methods, such as screw feeders delivering feedstock directly into the reactor bed, can pose mechanical and operational challenges. Temperature and pressure also play critical roles in fluidization dynamics by affecting interparticle forces and gas properties. Elevated temperatures lower gas density and increase viscosity, which reduces the minimum fluidization velocity $\left(U_{mf}\right)$ but may also cause agglomeration and defluidization in fine particles.

Despite their numerous advantages, BFBR applications have challenges related to feedstock variability, operational stability, and scalability. Irregularly shaped and low-density biomass particles can disrupt fluidization, leading to inefficiencies. Additionally, feeding systems in deep-bed configurations often encounter difficulties in achieving uniform distribution of feedstock and inert materials. Innovative reactor designs, such as top-fed systems and advanced distributor plates, aim to address these challenges by ensuring consistent fluidization and minimizing particle agglomeration. Such advancements are crucial for improving the scalability and commercial viability of BFBR technology.

This study expands on previous research by investigating the behavior of biomass in a top-fed, deep-bed BFBR, focusing on its hydrodynamics and thermochemical processes. It looks at important factors like bed shape, particle properties, and gas flow rates to improve pyrolysis and gasification processes. The goal is to better understand how BFBRs work and encourage their use in sustainable energy production. By addressing these challenges and taking advantage of the benefits of BFBRs, this research aims to promote the development of renewable energy technologies and contribute to a more sustainable energy future.

1.1 Bed pressure drop and minimum fluidization velocity

The pressure drop along the reactor bed height corresponds to the weight of particles per unit cross-sectional area. This term, known as the bed pressure drop, refers to the pressure drop across the entire bed when fluidized. The formula used to calculate this is:

$$\Delta P_b A = W, \text{where } W = mg = AH_{mf}\left(1 - \varepsilon_{mf}\right)\left(\rho_p - \rho_g\right)g \tag{1}$$

Here, ΔP_b is the bed pressure drop, A is the cross-sectional area, m is the bed's mass, g is the gravitational acceleration, H_{mf} is the bed height at minimum fluidization, ε_{mf} is the bed voidage at minimum fluidization, ρ_p is the particle density, and ρ_g is the fluidizing gas density. The bed voidage, ε, represents the ratio of void volume to bulk volume and can be calculated as:

$$\varepsilon_1 = 1 - \frac{\rho_b}{\rho_s} \tag{2}$$

Where ρ_b is the bulk density, and ρ_s is the skeletal density of the bed.

The minimum fluidization velocity (U_{mf}) is the gas velocity needed to initiate fluidization by balancing the bed weight. Theoretically, as gas velocity increases, the pressure drop per bed weight remains constant. Various correlations exist for predicting the minimum fluidization velocity, with Ergun's equation being a notable example:

$$\frac{\rho_g\left(\rho_p - \rho_g\right)\left(g d_p^3\right)}{\mu_g^2} = \frac{150\left(1 - \varepsilon_{mf}^2\right)}{\phi^2 \varepsilon_{mf}^3} \frac{\rho_g U_{mf} d_p}{\mu_g} + \frac{1.75}{\phi \varepsilon_{mf}^3} \frac{\rho_g^2 U_{mf}^2 d_p^2}{\mu_g^2} \tag{3}$$

where d_p is the mean particle diameter, μ_g fluidizing gas viscosity, and ϕ is the average particle sphericity.

Graphically, the minimum fluidization velocity can be determined by measuring the bed pressure drop as gas velocity increases. The intersection of slopes in fixed-bed and fluidization states indicates the minimum fluidization velocity. This process is illustrated in **Figure 1**.

1.2 Fluidization Regimes in Fluidized Bed Reactors

The fluidization regimes in a bed with increasing gas velocity (U_g) are depicted in **Figure 2**. Below the minimum fluidization velocity (U_{mf}), the bed remains in the "fixed bed" or "packed bed" state, as shown in **Figure 2a**. Initially, smaller particles are suspended, and eventually, all particles lift as the bed weight is balanced by the gas force. This creates the "minimum fluidization" condition illustrated in **Figure 2b**. As U_g increases further, bubbles form (**Figure 2c**), promoting better mixing and heat transfer. Larger bubble formations, or "slugs," appear as shown in **Figure 2d**, leading to rapid particle mixing (**Figure 2e**). A high gas velocity transitions the bed into fast fluidization (**Figure 2f**), and even higher velocities can transport particles out of the bed, known as pneumatic transport (**Figure 2g**).

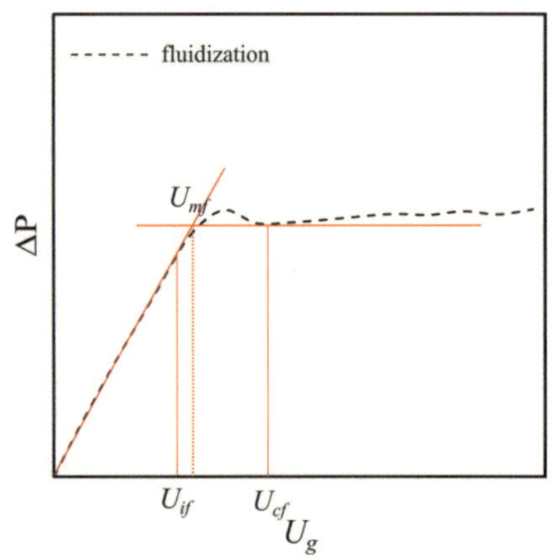

Figure 1.
Graphical solution to determine U_{if}, U_{mf}, and U_{cf} for increasing superficial gas velocity (fluidization).

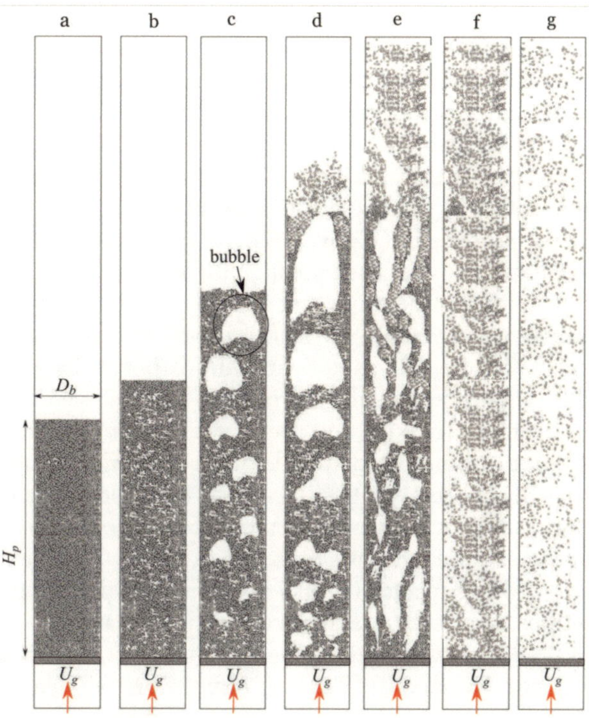

Figure 2.
Schematic illustration of the fluidization regimes g.

1.3 Geldart's particle classification

Fluidization behavior is influenced by particle diameter, sphericity, and density. Geldart [14] classified particles into four groups based on their fluidization characteristics:

Group A: Small-diameter (20–100 μm) and/or low-density particles, such as cracking catalysts, show dense phase expansion post-fluidization and require higher velocities for bubble formation compared to Group B.

Group B: Particles with diameters of 40–500 μm and densities of 1.4–4 g/cm^3, like sand, exhibit good fluidization and visible bubbles soon after minimum fluidization.

Group C: Highly cohesive particles with diameters of 10–80 μm mix and fluidize poorly due to strong interparticle forces. Bubbles appear shortly after fluidization with slight bed expansion.

Group D: Particles with diameters greater than 600 μm and high densities require higher gas flow rates to fluidize and show poorer fluidization behavior compared to Group B.

1.4 Gasification, pyrolysis, and bubbling fluidized bed reactor

Gasification is a thermochemical process that transforms carbon-based materials, such as coal, biomass, or waste, into a synthetic gas (syngas) by exposing them to high temperatures and a limited amount of oxygen (**Figure 3**). Through this controlled process, the feedstock undergoes chemical reactions that produce a gas mixture primarily composed of carbon monoxide (CO), hydrogen (H_2), and carbon dioxide

(CO_2). The resulting syngas can be used for a variety of purposes, such as generating electricity, producing fuels, or serving as a raw material for chemical processes. Gasification is considered a more efficient and environmentally friendly alternative to traditional combustion. A series of reactions that take place interactively in a gasification reaction are shown in **Table 1**.

On the other hand, pyrolysis is a crucial biomass conversion process that involves the thermal decomposition of organic materials at high temperatures (mainly from 300°C to 600°C) without oxygen (**Figure 3**). Pyrolysis methods include slow, rapid, ultrafast, and flash pyrolysis, with each yielding different product distributions based on process parameters like temperature and speed [1]. This process produces three primary products: biochar (a carbon-rich solid), bio-oil (a complex liquid), and syngas (a gaseous mixture of CO, H_2, CH_4, and light hydrocarbons). Pyrolysis includes primary reactions like dehydration, depolymerization, and fragmentation, as well as secondary reactions such as cracking and re-polymerization. Factors influencing product yield and composition include temperature, heating rate, residence time, and feedstock properties. As the temperature increases, the yields shift significantly. At temperatures above 700°C, pyrolysis favors the formation of syngas over biochar and bio-oil. The high-temperature pyrolysis reactions primarily involve (**Table 2**).

A bubbling fluidized bed reactor (BFBR) is a specific type of fluidized-bed reactor (FBR) where reactions take place due to the dynamic interaction between gas and solid particles in a fluidized state. Unlike fixed-bed reactors, which lack particle movement and mixing, fluidized-bed reactors offer improved heat transfer and

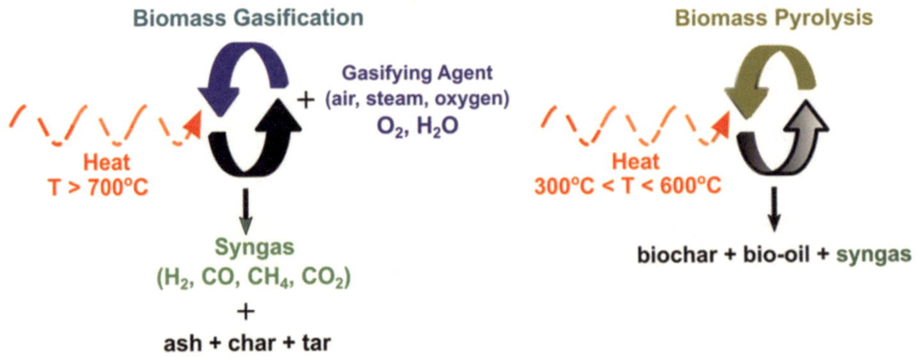

Figure 3.
Biomass gasification and pyrolysis reaction diagram.

Reaction name	Reaction formula	Enthalpy values
Water-gas shift reaction	$CO + H_2O \rightleftharpoons CO_2 + H_2$	$\Delta H^\circ 298K$ = -41 $kJmol^{-1}$
Steam methane reforming reaction	$CH_4 + H_2O \rightleftharpoons CO + 3H_2$	$\Delta H^\circ 298K$ = 206 $kJmol^{-1}$
Boudouard reaction	$C + CO_2 \rightleftharpoons 2CO$	$\Delta H^\circ 298K$ = 171 $kJmol^{-1}$
Methanation reaction	$C + 2H_2 \rightleftharpoons CH_4$	$\Delta H^\circ 298K$ = -75 $kJmol^{-1}$
Water-gas reaction	$C + H_2O \rightleftharpoons CO + H_2$	$\Delta H^\circ 298K$ = 131 $kJmol^{-1}$

Table 1.
Major 1st step gasification reactions.

Reaction name	Reaction formula
Char formation (Solid residue)	$\text{Biomass} \xrightarrow{\text{Heat}} \text{Biochar} + \text{Volatile Compounds}$
Steam methane reforming reaction	$CH_4 + H_2O \rightleftharpoons CO + 3H_2$
Volatile cracking	$\text{Volatiles} \xrightarrow{\text{Heat}} H_2 + CO + CO_2 + CH_4$
Secondary gas-phase reactions	$CH_4 + CO_2 \leftrightharpoons 2CO + 2H_2$

Table 2.
Major pyrolysis reactions at high temperatures.

carbon conversion efficiency, making them ideal for advanced thermochemical processes [15, 16]. Fluidization refers to the condition where solid particles are suspended and moved by the upward flow of a fluidizing medium, such as gas or liquid, inside a vertical column (**Figure 4**).

The effectiveness of mixing and heat transfer in a BFBR is significantly influenced by bubble dynamics within the fluidized bed. Alongside bubble behavior, the physical and thermal properties of the bed material are critical. Factors like particle size, shape, density, and thermal conductivity directly impact fluidization dynamics, heat transfer efficiency, and the rate of reaction. The term "bed" refers to the mixture of solid particles and fluidizing agents in the reactor. **Figure 4a** and **b** depict a typical fluidized-bed reactor and a 3D CAD model, respectively.

Key components of a BFBR include a plenum for introducing the fluidizing medium, a distributor plate to ensure even flow and maintain the pressure drop required for bubble formation, the reactor bed where the reactions occur, and a freeboard that reduces gas velocity to allow entrained particles to settle back into the bed.

During operation, the feedstock is typically introduced into the reactor through a screw feeder, which can add material from various locations, including the bottom,

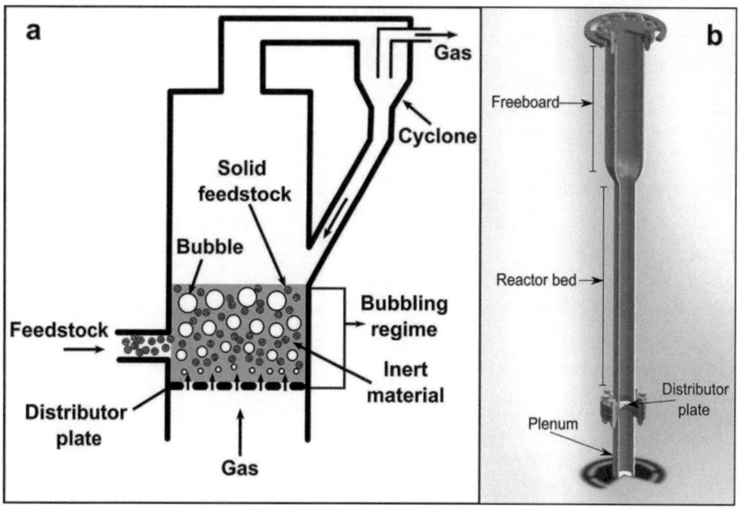

Figure 4.
(a) Fluidized-bed reactor diagram and (b) 3D CAD model.

sides, or top of the reactor. The reaction process starts when the carbonaceous feedstock is exposed to the reactor bed at the appropriate temperature. Heavier byproducts, such as ash, settle at the bottom of the bed, while lighter particles, like char, may be carried out by the gas flow. In systems designed for high efficiency, such as circulating fluidized bed (CFB) reactors, cyclone separators are used to capture and return these light particles to the bed. Compared to CFB reactors, BFBRs are more cost-effective and versatile, able to handle a wider range of feedstocks, making them suitable for various industrial applications [17]. After gasification or pyrolysis, the syngas produced is cooled and filtered to remove impurities before being utilized in downstream processes.

2. Experimental setup and methodology

2.1 Cold flow and BFBR test rigs

The experimental system comprises three main test rigs: the cold flow rig (**Figure 5a**), the bubbling fluidized bed reactor (BFBR) rig (**Figure 5b**), and a fixed-bed reactor. The cold flow setup, constructed from transparent acrylic, is designed for visualizing fluidization patterns and measuring pressure drops under ambient conditions. Key components of the cold flow rig include a plenum, a distributor plate, the

Figure 5.
a) Cold flow and b) BFBR experimental setups.

Material	Polymethyl-methacrylate
Reactor internal diameter (cm)	3.81
Reactor tube height (cm)	40.64
Transition cone height (cm)	2.54
Freeboard internal diameter (cm)	7.62
Freeboard height (cm)	25.4

Table 3.
Cold flow test rig characteristics.

bed section, and the freeboard, with detailed specifications outlined in **Table 3**. A stainless steel distributor plate featuring a 10-μm pore size and 39% porosity was utilized. A mass flow controller regulates the flow of the fluidizing gases, such as nitrogen or air.

Pressure measurements were taken at nine vertically aligned points, spaced 3.81 cm apart, from just below the distributor plate (point 1) to the region just beneath the transition cone (point 9). These pressure readings were recorded at a 10 Hz sampling rate using a data acquisition system and later processed with Python-based analytical tools. **Figure 5a** illustrates the positions of the pressure taps along the reactor. High-speed cameras captured fluidization behavior, and the images were enhanced using Python-Scikit, an open-source image processing software, to improve contrast and visualize material mixing more effectively.

Cold flow experiments were performed using total mixture masses of 100 g, 200 g, and 300 g, with coal feedstock constituting 4% of the total weight and placed on top in a segregated state to replicate the hydrodynamics of BFBR at elevated temperatures. Measurements for each case were recorded after a stabilization period of at least 30 seconds to ensure consistency. Tests were conducted three times, and no significant differences were observed.

The BFBR test stand incorporates a reactor, a top-load furnace, a screw feeder, and a micro gas chromatograph for analyzing the composition of syngas produced during gasification tests. The reactor, housed within a furnace capable of reaching 1500°C, was constructed from Inconel steel with dimensions mirroring those of the cold flow rig. Temperatures inside the reactor, above the inert material, and along the walls were monitored using K-type thermocouples.

Pressure drop measurements in the BFBR rig assessed the impact of elevated temperatures on bed hydrodynamics. Experiments were performed using unary sand mixtures weighing 200 g and 300 g, with identical durations to those of the cold flow tests. For further details on the design and configuration of these test stands, including fixed-bed reactors, readers are referred to the work of Sivri [18].

2.2 Material analysis and preparation

In this research, hardwood sawdust was utilized as the feedstock, while sand served as the bed material for both cold flow and actual BFBR experiments. Commercial-grade fine silica sand from Quikrete brands was specifically chosen for its properties as the bed material. The moisture content, volatile matter, ash, and elemental composition of the hardwood sawdust were analyzed, with the results

	Carbon (C), (%)	Hydrogen (H), (%)	Oxygen (O), (%)	Sulfur (S), (%)	Moisture (%)	Ash (%)
Hardwood	45.25	4.65	49.2	0	7.16	0.32

Table 4.
Elemental and proximate analysis (by mass) of hardwood.

Material	Average of sphericity	Sauter mean diameter (μm)	Bulk density (g/cm^3)	Skeletal density (g/cm^3)	Geldart's group
Sand	0.86	324	1.43	2.64	B
Hardwood	0.56	468	0.27	1.47	B
Hardwood (wt%4) and Sand	0.85	329	1.21	2.59	B

Table 5.
Material size, density, sphericity, and Geldart's group analysis.

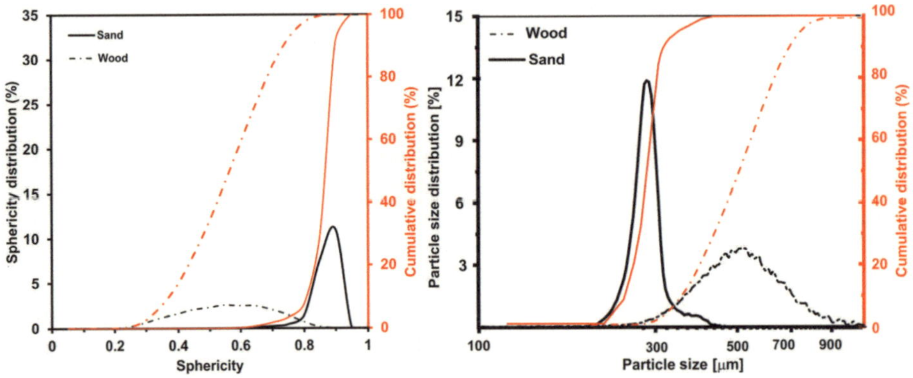

Figure 6.
Particle size and sphericity distributions.

presented in **Table 4**. Additionally, particle size and sphericity distributions were examined using a dynamic image analysis system (Sympatec GmbH, Model QICPIC), and skeletal densities were determined with a gas pycnometer (AccuPyc, Model 1330 Helium Pycnometer). **Table 5** summarizes the outcomes of these analyses for both the feedstock and inert materials.

The sand exhibited a narrower distribution in both sphericity (90% between 0.85 and 0.95) and particle size (90% between 235 μm and 347 μm) compared to the sawdust particles (**Figure 6**). In contrast, hardwood sawdust had a larger mean diameter of 468 μm compared to the sand's 324 μm. Sand and hardwood mixture contributed to improved packing and enhanced fluidization characteristics when compared to hardwood-only bed packaging and fluidization quality.

3. Fluidization analyses

This section presents the fluidization analysis results for a sand-sawdust mixture (sawdust accounted for 4 wt% of the total mixture) at ambient temperature and for sand alone at an elevated temperature of approximately 800°C. Deep-bed applications offer advantages such as improved reaction efficiency and enhanced heat transfer rates. However, achieving a quasi-homogeneous mixture, particularly in binary systems where biomass is mixed with bed material, remains a challenge, especially in top-fed or on-bed feeding configurations. To address this, fluidization behavior was analyzed at three different bed heights while incrementally increasing the gas velocity. Key parameters, including bed pressure drop, minimum fluidization velocity, and mixing behavior, were examined. Additionally, high-speed imaging was employed to capture bed behavior across varying gas velocities.

To further investigate these dynamics, experiments were conducted using mixtures with total masses of 100 g, 200 g, and 300 g. Sawdust accounted for 4% of the total mass at the initial condition, and it was placed on top of the bed. The bed pressure drop (ΔP_b) was measured as the gas velocity increased in increments of 0.0146 m/s. The minimum fluidization velocity (U_{mf}) was determined from the bed pressure drop behavior. Additionally, high-speed imaging was used to capture the fluidization and mixing dynamics at each gas flow rate after fluidization was achieved.

In **Figures 7–9**, the column images beneath the pressure drop curves illustrate the mixture behavior at each gas velocity. The images were used to estimate the optimal gas velocity range for achieving an almost homogeneous (quasi-homogeneous) binary mixture with stable fluidization characteristics suitable for BFBR operation. The mixture of sand and sawdust exhibited a higher bulk density (1.21 g/cm^3 compared to 0.27 g/cm^3) and greater average sphericity (0.85 versus 0.56) (**Table 5**) than the unary sawdust mixture, which is based on the particle characteristics of sawdust alone. Employing an inert bed material with improved particle properties, such as higher sphericity and density, resulted in enhanced fluidization behavior and, consequently, better mixing performance compared to the sawdust-only mixture. The variation in bed pressure drop and mixture behavior of the binary mixture with a total mass of 100 grams as a function of increasing gas velocity is presented in **Figure 7**. The bed aspect ratio was approximately 2 at the steady state (fixed bed state). During the fixed bed state, defined as the gas velocity interval where the bed pressure drop increases linearly, the particles were stationary. However, with the further increase in the gas velocity, small sand particles percolated into the sawdust layer. Preferential channel formations were observed near the upper region of the sand layer, with a peak pressure drop of approximately $\Delta P_b = 0.8$ kPa. As the fluidization process progressed, a minor pressure drop of around 0.05 kPa occurred during the transition from minimum fluidization to complete fluidization, following the linear increase in the pressure curve leading up to fluidization onset. The minimum fluidization velocity, U_{mf}, was observed to be around 0.04 m/s. Before reaching complete fluidization, tiny bubbles associated with channeling were visible at gas velocities between 0.075 m/s and 0.1 m/s, as shown in **Figure 7b** columns 3–4). Bubbling regime started with the gas velocity increment, and around $U_g = 0.13$ m/s, a well-mixed state was achieved. At higher gas velocities above 0.16 m/s, slug formations were close to flat formations, as seen in (**Figure 7b** columns 9–11).

The pressure drop change for the 200 g mixture is shown in **Figure 8**. The initial bed aspect ratio was approximately 4. The pressure drop increased almost linearly until reaching fluidization. Compared to the 100 g mixture mass case, the transition to complete fluidization was sharper, with a pressure drop of $\approx$0.4 kPa. The minimum fluidization velocity, U_{mf}, was observed to be around 0.07 m/s. Furthermore, mixing began at $U_g \approx 0.12$ m/s, just after reaching complete fluidization, with the penetration

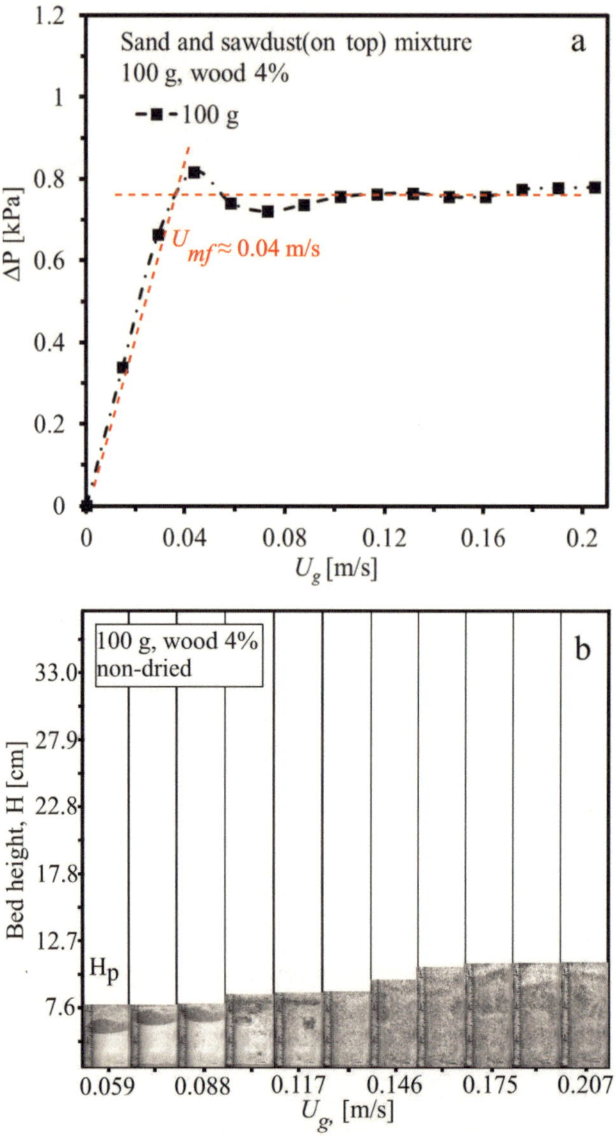

Figure 7.
Bed pressure drop and fluidization behavior for 100 g sawdust and sand mixture with fluidizing gas velocity.

of sand particles into the sawdust layer (**Figure 8b** column 5). Slug formations were visible at $U_g \approx 0.13$ m/s (**Figure 7b** columns 9–11). Gas velocity was insufficient to overcome the interparticle forces within the sawdust layer, thus preventing the disruption of its unity until $U_g \approx 0.16$. A quasi-homogeneous mixture was only achieved at gas velocities above $U_g \approx 0.16$, at which flat slug formations were dominant.

Subsequent experiments were carried out to examine the fluidization behavior of a deeper bed with an aspect ratio of ≈4 and a total mass of 300 g, with sawdust constituting 4% of the total mass. The fluidization behavior and pressure drop trends for the 300 g bed, with increasing gas velocity, are presented in **Figure 9**. Similar to

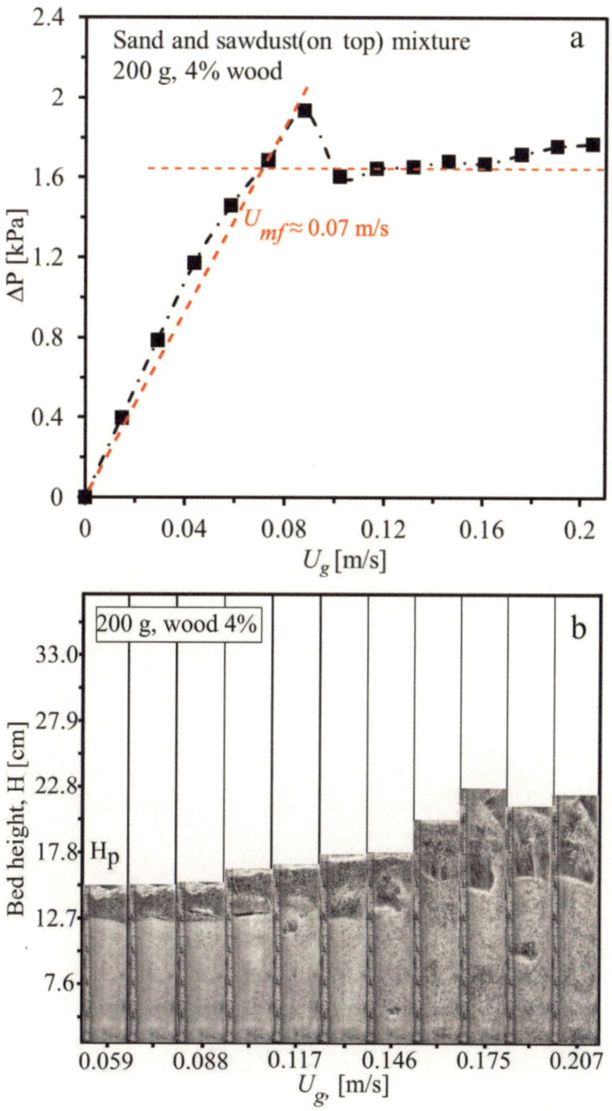

Figure 8.
Bed pressure drop and fluidization behavior for 200 g sawdust and sand mixture with fluidizing gas velocity.

earlier tests with 100 g and 200 g total masses, the bed pressure drop showed a linear increment during the fixed-bed state. The highest pressure drop reached around 3 kPa, and the transition from fixed bed to fluidization occurred at approximately 0.6 kPa. The bed achieved complete fluidization at a gas velocity of about 0.1 m/s. For gas velocities between 0.1 m/s and 0.13 m/s (**Figure 9b**, columns 4–6), the sawdust layer started getting apart from the sand layer. At a gas velocity of approximately 0.14 m/s (**Figure 9b**, column 7), the sawdust layer was completely separated and elevated further from the sand. Friction and interparticle forces were capable of holding the layer as a whole. An increment in gas velocity caused the sawdust layer to rise further due to the dominant flat slug formation in the bed. Although a small

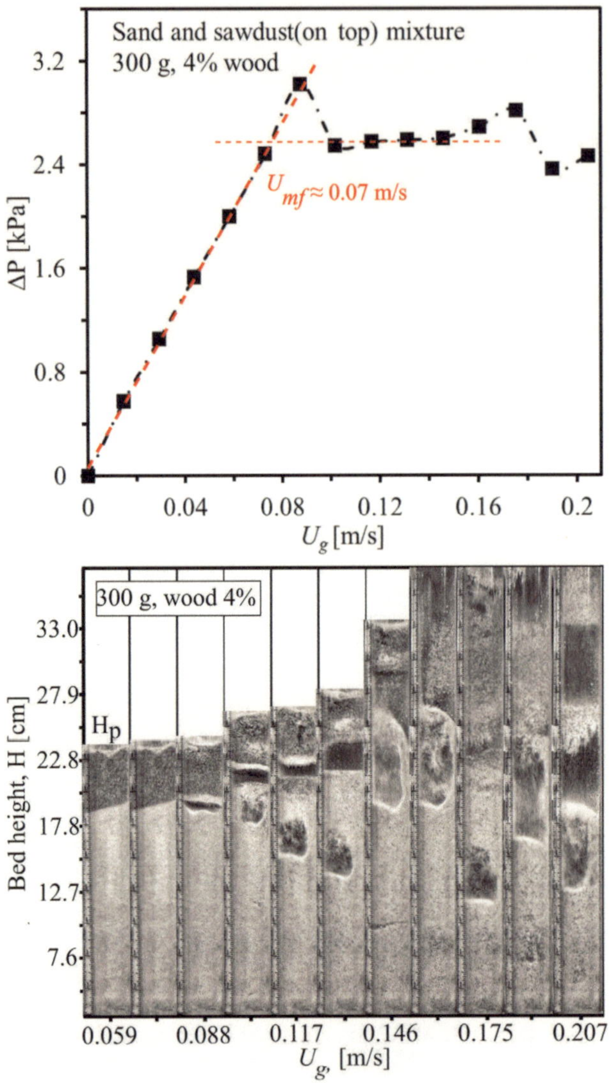

Figure 9.
Bed pressure drop and fluidization behaviors with increasing superficial gas velocity for the mixtures of sand and sawdust with a total mixture mass of 300 g.

portion of the layer mixed with the bed, the sawdust layer largely retained its form. Based on observations, when the bed aspect ratio is 6, the BFBR does not ensure proper mixing, making this aspect ratio unsuitable for operation.

Sawdust biomass particles are highly reactive and exhibit immediate changes in their characteristics during reactions. To better understand the bed behavior without the presence of sawdust, experiments were conducted with sand material only at around 800°C, using total masses of 200 g and 300 g. The bed pressure drop was studied for unary sand mixtures at elevated temperatures ranging from 800°C to 805°C (**Figure 10**). In the cold flow test rig, which had similar dimensions to the BFBR, the bed aspect ratios were measured at 3.3 and 4.83 for the 200 g and 300 g total masses, respectively. The bed pressure drop exhibited a linear increase at all

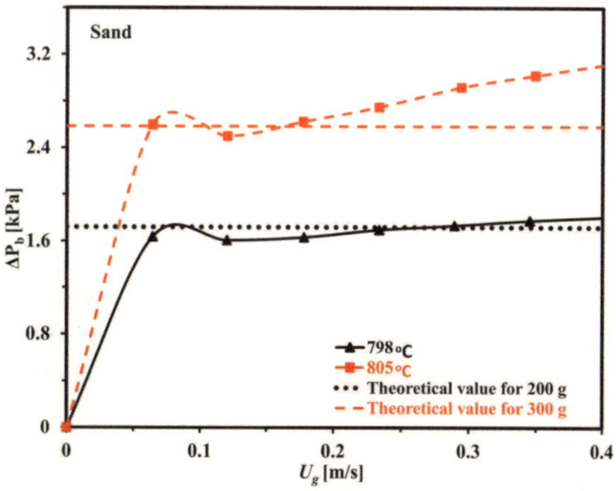

Figure 10.
Bed pressure drop versus fluidization gas velocity around 800°C.

temperatures for both total bed masses, followed by a gradual transition to complete fluidization in each case. The bed pressure drop exhibited a linear increase at all temperatures for both total bed masses, transitioning gradually to complete fluidization in each case. The minimum fluidization velocity, U_{mf}, was found to be approximately 0.07 m/s for both bed aspect ratios, supporting previous studies that observed no significant change in U_{mf} across different aspect ratios. After achieving complete fluidization, the pressure drop continued to rise with increasing gas velocity due to three primary factors: wall effects [19, 20], increased bed voidage [21], and stronger interparticle forces, especially for this narrow-sized (3.81 cm diameter) deep-bed reactor. As expected, the bed pressure drop increased more steeply for the bed with the higher aspect ratio, owing to stronger wall effects.

3.1 Summary and conclusions

The fluidization and mixing behaviors of sand and sawdust mixtures were investigated at different bed aspect ratios, with total masses of 100 g, 200 g, and 300 g, and sawdust making up 4% of the total mass. Initially, the mixtures were in a segregated state, with sawdust placed on top of the bed material to simulate top-fed feeding. As the gas velocity increased, the fluidization and mixing behaviors of the mixtures were observed. The following key conclusions were drawn:

- Beds with initial aspect ratios of ≈2 and ≈4 were able to achieve well-mixed, quasi-homogeneous fluidization. However, beds with aspect ratios greater than 4 did not achieve effective mixing, indicating that fluidized bed operations with aspect ratios above 4 are not recommended. Additionally, beds with lower aspect ratios (≤ 2) reached quasi-homogeneous mixing at lower gas velocities compared to those with higher aspect ratios (≥ 2). Bed with initial bed aspect ratios of ≈2 and ≈4 could reach well-mixed (quasi-homogenous mixing), while aspect ratios larger than 4 could not reach.

- At higher gas velocities, flat slug formations were observed across all bed aspect ratios, disrupting the uniformity of mixing and fluidization. Moreover, higher bed aspect ratios (≥ 4) prevented effective fluidization and mixing due to the inability to break the strong interparticle forces in the sawdust layer at the top of the bed, leading to substantial elutriation of sawdust particles.

- Unary sand mixtures showed stable fluidization behavior at elevated temperatures, with wall effects and increased bed voidage playing a significant role in the pressure drop behavior.

- Finally, due to the high reactivity and irregular shape of biomass particles, it is recommended that the feedstock be mixed immediately with the bed material to ensure optimal reaction efficiency and effective fluidization.

4. Reaction analysis

Pyrolysis, the thermal decomposition of organic materials in the absence of oxygen, typically occurs at reaction temperatures between 300^oC and 600^oC. It is a critical process for converting biomass into valuable products such as gases, bio-oil, and char. This subsection presents the results of rapid pyrolysis conducted at $\approx 900^oC$ in a bubbling fluidized-bed reactor (BFBR) and slow pyrolysis performed in a downdraft fixed-bed reactor, comparing the outcomes to findings from the literature.

The gas compositions obtained from both reactor types operating at 900^oC were analyzed and compared. The choice of reactor significantly influenced product distribution and gas composition. Hardwood, the biomass feedstock used in this study, had an elemental composition of approximately 50% oxygen and 7.16% moisture, as shown in **Table 4**. Nitrogen served as both the fluidizing agent and transport gas, while sand was selected as the bed material due to its low thermal conductivity, which enhances thermal management within the reactor.

The BFBR experiments were conducted in a slightly fast fluidization regime to ensure rapid mixing and improved particle-particle and particle-gas interactions. This configuration provided efficient heat transfer between the bed material (sand) and the sawdust particles, optimizing reaction kinetics. The interplay between reaction kinetics and fluidization hydrodynamics, including bubble formation, frequency, and size, plays a critical role in determining the efficiency of pyrolysis. To achieve consistent and effective pyrolysis, the bed medium was preheated to 900^oC, and temperature measurements were taken at the reactor wall, within the bed material, and just above the bed.

In each experiment, 2 grams of sawdust were fed into the reactor through a supply line using pressurized nitrogen at a rate of 2 g/min. For the fixed-bed reactor, 1 gram of sawdust was heated at a rate of $20^oC/\,min$ and held at 900^oC for 30 minutes. Reactor pressure varied with temperature, with values around 250 psig at 750^oC, 380 psig at 850^oC, and 420 psig at 900^oC.

The gas composition from the fixed-bed reactor was H_2 (19%), CO (55%), CO_2 (12%), and CH_4 (13%), excluding the inert gas content. These results align with typical trends observed in fixed-bed pyrolysis, where limited heat transfer and extended residence times favor char formation and higher methane yields. The dominance of CO is consistent with the thermal decomposition of cellulose and hemicellulose, which release significant amounts of CO and H_2, while lignin decomposition contributes to CH_4 production.

Studies suggest that CO concentrations in wood pyrolysis reactions typically range from 45 to 60%, while H_2 levels vary between 15% and 35%, depending on factors such as feedstock composition, particle size, reaction conditions, and reactor design [4, 6, 22–25]. The results of this study fall within these ranges, demonstrating the ability of both reactor types to achieve effective thermal decomposition under high-temperature conditions. Comparative gas compositions from the literature are summarized in **Table 6**.

In the BFBR, the gas composition was H_2 (38%), CO (42%), CO_2 (9%), and CH_4 (9%). The higher H_2 yield in the BFBR highlights its efficiency in promoting secondary gas-phase reactions, such as the water-gas shift and methane reforming. Improved heat transfer and better mixing in fluidized beds facilitate these reactions, leading to lower CH_4 and CO_2 yields compared to the fixed-bed reactor. Syngas compositions obtained during slow pyrolysis in the fixed-bed reactor and fast pyrolysis in the BFBR are shown in **Figure 11**.

The literature consistently reports that BFBRs outperform fixed beds in maximizing gas yields due to their uniform temperature distribution and enhanced gas-solid contact. The findings of this study are consistent with these observations, emphasizing the suitability of the BFBR for hydrogen-rich syngas production. The higher H_2 yield in the BFBR is attributed to efficient gas-solid interactions and secondary reactions. These trends are supported by studies such as [26, 27], which also observed enhanced gas yields and hydrogen production at high temperatures.

Increasing temperature and residence time have been shown to favor cracking reactions of hydrocarbons, resulting in higher gas yields and H_2 composition [2, 4, 22, 28–30]. Both reactors produced CO as the dominant gas, but the fixed-bed reactor exhibited higher CH_4 levels due to less effective cracking and reforming. The lower CO_2 and CH_4 yields in the BFBR reflect improved gas-phase reaction efficiencies, consistent with findings from [28]. While the fixed-bed reactor is more suited for char and tar production, the BFBR is ideal for maximizing gas yields, particularly hydrogen. The fixed bed's limited heat transfer reduces its efficiency compared to the fluidized bed [29].

Authors	***T*** **(°C)**	**Reactor type**	**Pyrolysis type**	**H_2 (%)**	**CO (%)**	**CO_2 (%)**	**CH_4 (%)**	**Other hydrocarbons (%)**
Chen et al. [22]	950	Entrained flow	Fast	24	50.7	6.5	10	18
Commandré et al. [23]	900	Entrained flow	Fast	21	56.8	3.9	11	7.3
Hemati et al. [24]	900	Fluidized bed	Fast	18.5	56	7.5	12.5	5
Mok et al. [25]	900	Entrained flow	Ultra	22.9	57.7	3.5	12	15.9
Zanzi [4]	800; 1000	Fixed bed	Fast	17; 34	51; 46	8.3; 7.5		23; 12
Waheed et al. [6]	850	Fixed bed	Slow; fast	20.7; 24	39.3; 48	32.1; 10.8	6.7; 11.8	1.3; 55

Table 6.
Comparison of the volumetric gas composition from wood pyrolysis with the literature.

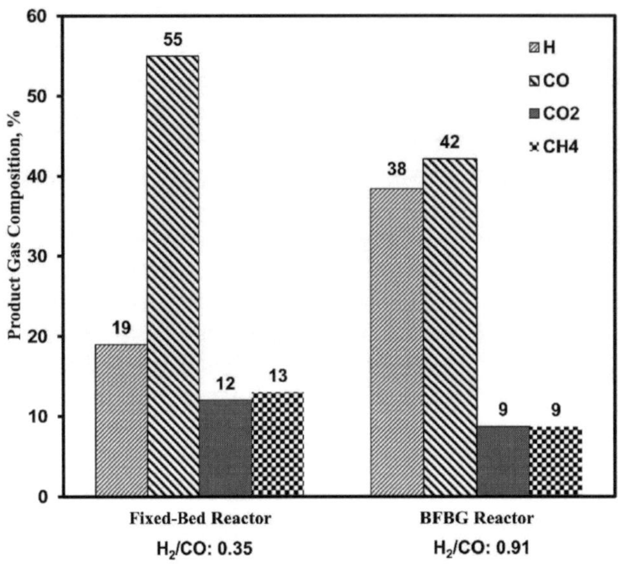

Figure 11.
Syngas compositions obtained during the reaction at different reactor types at 900 °C.

The observed trends emphasize the critical role of reactor type and operating conditions in determining gas composition during wood pyrolysis. The BFBR's high hydrogen yield shows its potential for applications targeting hydrogen-rich syngas production. However, further optimization of residence time and temperature could improve syngas quality by increasing CO yield and reducing CO_2. When compared with the literature, the results of this study align well with reported trends while highlighting the variability introduced by reactor design and operational parameters.

4.1 Summary and conclusions

The gas compositions and reaction dynamics of hardwood pyrolysis in bubbling fluidized-bed and fixed-bed reactors at 900°C were investigated and compared with literature findings. This study analyzed the influence of reactor type, operating conditions, and reaction temperature on gas composition. The following key conclusions were drawn:

- The bubbling fluidized-bed reactor (BFBR) demonstrated significantly higher hydrogen production (H_2: 38%) compared to the fixed-bed reactor (H_2: 19%), primarily due to enhanced secondary gas-phase reactions, better heat transfer, and improved mixing. This highlights the BFBR's superior performance for hydrogen-rich syngas production.

- Carbon monoxide (CO) was the primary gas in both reactors, with concentrations of 55% in the fixed bed and 42% in the BFBR. The predominance of CO is attributed to the thermal decomposition of cellulose and hemicellulose.

- Methane (CH_4) and carbon dioxide (CO_2) concentrations were higher in the fixed-bed reactor (CH_4: 13%, CO_2: 12%) compared to the BFBR (CH_4: 9%, CO_2: 9%), reflecting the limited reforming and cracking efficiency of the fixed bed.

- The BFBR exhibited lower CH_4 and CO_2 yields due to improved gas-phase reaction efficiencies, including the water-gas shift and methane reforming reactions, facilitated by better heat and mass transfer.

- Higher temperatures and shorter residence times in the BFBR favored gas formation over char and tar, aligning with trends observed in the literature. Optimization of operating parameters, such as residence time and catalyst addition, could further enhance syngas quality.

- While the fixed-bed reactor is more suitable for applications focusing on char and tar production, the BFBR is optimal for maximizing gas yields, particularly for hydrogen-rich syngas applications.

In conclusion, this study emphasizes the critical role of reactor design and operating conditions in determining the gas composition during wood pyrolysis. The BFBR's high hydrogen yield and superior gas-phase reaction efficiency make it highly suitable for processes targeting hydrogen-rich syngas production. Future work should explore the effects of temperature, residence time, and catalyst addition to further improve gas composition and maximize hydrogen production.

Acknowledgements

The research was conducted at the WVU's Advanced Combustion Laboratory in Morgantown, WV, and the author gratefully acknowledges WVU's CIGRU, MAE, and CBE departments for their assistance with comprehensive mechanical, electronic, and data acquisition hardware/software systems, respectively. This work was partially supported by the US Department of Energy's Fossil Energy Advanced Gasification Program and the NETL Research and Innovation Center's Advanced Gasification effort within the Advanced Reaction Systems FWP.

Author details

Ali Can Sivri
Department of Mechanical and Aerospace Engineering, West Virginia University, Morgantown, USA

*Address all correspondence to: acsengresearch@gmail.com

IntechOpen

© 2025 The Author(s). Licensee IntechOpen. This chapter is distributed under the terms of the Creative Commons Attribution License (http://creativecommons.org/licenses/by/4.0), which permits unrestricted use, distribution, and reproduction in any medium, provided the original work is properly cited.

References

[1] Uddin MN, Techato K, Taweekun J, Rahman MM, Rasul MG, Mahlia TMI, et al. An overview of recent developments in biomass pyrolysis technologies. Energies. 2018;**11**(11):3115. DOI: 10.3390/en11113115

[2] Zanzi R, Sjöström K, Björnbom E. Rapid pyrolysis of wood with application to gasification. In: Advances in Thermochemical Biomass Conversion. Netherlands: Springer; 1993. pp. 977-985. DOI: 10.1007/978-94-011-1336-6_75

[3] Wan Sulaiman WR, Lee ES. Pyrolysis of eucalyptus wood in a fluidized-bed reactor. Research on Chemical Intermediates. 2012;**38**(8):2025-2039. DOI: 10.1007/s11164-012-0523-z

[4] Zanzi R. Pyrolysis of biomass. Rapid pyrolysis at high temperature. Slow pyrolysis for active carbon preparation [PhD thesis]. KTH, Chemical Engineering and Technology; 2001;**144**: 52. ISSN: 1104-3466

[5] Zhang H, Xiao R, Wang D, He G, Shao S, Zhang J, et al. Biomass fast pyrolysis in a fluidized bed reactor under N_2, CO_2, CO, CH_4 and H_2 atmospheres. Bioresource Technology. 2011;**102**(5): 4258-4264. DOI: 10.1016/j.biortech.2010.12.075

[6] Waheed QMK, Nahil MA, Williams PT. Pyrolysis of waste biomass: Investigation of fast pyrolysis and slow pyrolysis process conditions on product yield and gas composition. Journal of the Energy Institute. 2013;**86**(4):233-241. DOI: 10.1179/1743967113z.00000000067

[7] Liu M, Zhang HQ, Chan CK, Lau KS, Lin WY. Study of cluster formation in dense two-phase flow using a multi-lattice deterministic model. Powder Technology. 2006;**162**(3):175-182. DOI: 10.1016/j.powtec.2005.12.006

[8] Zhou L, Zhao Z, Li W, Bai L, Agarwal RK. Influence mechanism of particle density in a gas-solid fluidized bed. Physics of Fluids. 2023;**35**(12): 123345. DOI: 10.1063/5.0180227

[9] Si C, Guo Q. Fluidization characteristics of binary mixtures of biomass and quartz sand in an acoustic fluidized bed. Industrial & Engineering Chemistry Research. 2008;**47**(23): 9773-9782. DOI: 10.1021/ie801070z

[10] Sun G, Grace JR. Effect of particle size distribution in different fluidization regimes. AICHE Journal. 1992;**38**(5): 716-722. DOI: 10.1002/aic.690380508

[11] Amiri Z, Movahedirad S. Bubble-induced particle mixing in a 2-D gas-solid fluidized bed with different bed aspect ratios: A CFD-DPM study. Powder Technology. 2017;**320**:637-645. DOI: 10.1016/j.powtec.2017.07.097

[12] Formisani B, Girimonte R, Longo T. The fluidization pattern of density-segregating binary mixtures. Chemical Engineering Research and Design. 2008; **86**(4):344-348. DOI: 10.1016/j.cherd.2007.11.004

[13] Wilk V, Schmid JC, Hofbauer H. Influence of fuel feeding positions on gasification in dual fluidized bed gasifiers. Biomass and Bioenergy. 2013; **54**:46-58. DOI: 10.1016/j.biombioe.2013.03.018

[14] Cranfield RR, Geldart D. Large particle fluidisation. Chemical Engineering Science. 1974;**29**(4): 935-947. DOI: 10.1016/0009-2509(74)80085-0

[15] Lim MT, Alimuddin Z. Bubbling fluidized bed biomass gasification—Performance, process findings and energy analysis. Renewable Energy. 2008;**33**(10):2339-2343. DOI: 10.1016/j.renene.2008.01.014

[16] Narváez I, Orío A, Aznar MP, Corella J. Biomass gasification with air in an atmospheric bubbling fluidized bed. Effect of six operational variables on the quality of the produced raw gas. Industrial & Engineering Chemistry Research. 1996;**35**(7):2110-2120. DOI: 10.1021/ie9507540

[17] Cui H, Grace JR. Fluidization of biomass particles: A review of experimental multiphase flow aspects. Chemical Engineering Science. 2007;**62** (1–2):45-55. DOI: 10.1016/j.ces.2006.08.006

[18] Sivri AC. A study on the hydrodynamics of a bench-scale top-fed bubbling fluidized bed gasifier using biomass and coal as feedstocks [PhD thesis]. West Virginia University Libraries. 10.33915/etd.7907

[19] Srivastava A, Sundaresan S. Role of wall friction in fluidization and standpipe flow. Powder Technology. 2002;**124**(1–2):45-54. DOI: 10.1016/s0032-5910(01)00471-5

[20] Olatunde G, Fasina O, McDonald T, Adhikari S, Duke S. Moisture effect on fluidization behavior of loblolly pine wood grinds. Biomass Conversion and Biorefinery. 2016;7(2):207-220. DOI: 10.1007/s13399-016-0223-9

[21] Formisani B, Girimonte R, Mancuso L. Analysis of the fluidization process of particle beds at high temperature. Chemical Engineering Science. 1998;**53**(5):951-961. DOI: 10.1016/s0009-2509(97)00370-9

[22] Chen L, Dupont C, Salvador S, Boissonnet G, Schweich D. Influence of particle size, reactor temperature and gas phase reactions on fast pyrolysis of beech wood. International Journal of Chemical Reactor Engineering. 2010;**8**(1):1-19. DOI: 10.2202/1542-6580.1922

[23] Commandré J-M, Lahmidi H, Salvador S, Dupassieux N. Pyrolysis of wood at high temperature: The influence of experimental parameters on gaseous products. Fuel Processing Technology. 2011;**92**(5):837-844. DOI: 10.1016/j.fuproc.2010.07.009

[24] Hemati M, El Ghezal L, Laguerie C. Etude expérimentale de la pyrolyse de sciure de bois dans un lit fluidisé de sable entre 630 et 940 °C. The Chemical Engineering Journal. 1989;**42**(2):B25-B38. DOI: 10.1016/0300-9467(89)85009-9

[25] Mok LK, Graham RG, Freel BA, Bergougnou MA, Overend RP. Fast pyrolysis (ultrapyrolysis) of cellulose and wood components. Journal of Analytical and Applied Pyrolysis. 1985;**8**: 391-400. DOI: 10.1016/0165-2370(85)80038-3

[26] Lan P, Xu Q, Zhou M, Lan L, Zhang S, Yan Y. Catalytic steam reforming of fast pyrolysis bio-oil in fixed bed and fluidized bed reactors. Chemical Engineering & Technology. 2010;**33**(12):2021-2028. DOI: 10.1002/ceat.201000169

[27] Fernandez E, Amutio M, Artetxe M, Arregi A, Santamaria L, Lopez G, et al. Assessment of product yields and catalyst deactivation in fixed and fluidized bed reactors in the steam reforming of biomass pyrolysis volatiles. Process Safety and Environmental Protection. 2021;**145**:52-62. DOI: 10.1016/j.psep.2020.07.039

[28] Santamaria L, Beirow M, Mangold F, Lopez G, Olazar M, Schmid M, et al. Influence of temperature on products from fluidized bed pyrolysis of wood and solid recovered fuel. Fuel. 2021;**283**: 118922. DOI: 10.1016/j.fuel.2020.118922

[29] Patel S, Kundu S, Halder P, Veluswamy G, Pramanik B, Paz-Ferreiro J, et al. Slow pyrolysis of biosolids in a bubbling fluidised bed reactor using biochar, activated char and lime. Journal of Analytical and Applied Pyrolysis. 2019;**144**:104697. DOI: 10.1016/j.jaap.2019.104697

[30] Oyebanji JA, Oyedepo SO. Compositional and experimental investigation of the effect of reactor temperature on softwood and hardwood pyrolysis. World Journal of Engineering. 2018;**15**(1):21-26. DOI: 10.1108/wje-03-2017-0053

Chapter 7

Biomedical Waste Plasma Gasification: A Case Study of Brazil

Regina Franciélle Silva Paulino,
Alexei Mikhailovich Essiptchouk and José Luz Silveira

Abstract

Plasma gasification technology is one of the environmentally friendly techniques that can be applied to process municipal, commercial, industrial, petrochemical and biomedical waste. One of the main challenges in using this technology is the high electricity consumption and the high initial investment cost, but for biomedical waste, it can be an alternative due to the difficulty and high economic cost of its disposal, producing ash and syngas that can be used to generate thermal or electrical energy. Considering that Brazil has serious problems with incorrect disposal of biomedical waste (BW), this book chapter aims to present thermodynamic studies carried out to determine the potential of electric energy generation using the syngas produced in a plasma gasification system in an internal combustion engine and in a gas turbine set, and economic engineering studies are also carried out to determine the syngas and electricity production cost in US$/kWh, annual saving expected and payback. Finally, thermodynamic and economic studies are applied to the case study considering the processing of biomedical waste from the health service produced in the São Paulo city, São Paulo state, Brazil.

Keywords: plasma gasification, biomedical waste, syngas, thermodynamic analysis, economic analysis, case study

1. Introduction

1.1 Biomedical waste

Biomedical waste (BW) comes from activities in health service units, dental, medical, pharmaceutical and laboratory clinics, and teaching and research centers in the areas of both human and animal health, including blood, expired vaccines, blood products, organic fluids, tissues, radioactive waste, sharps, chemical and pharmaceutical waste, and others [1].

According to the latest report presented in the Panorama of Solid Waste in Brazil, base year 2022, carried out by ABREMA [2], it was pointed out that Brazil generated 307 thousand tons of biomedical waste and approximately 80% of Brazilian municipalities declared that these BW were collected, with the main destination being the incineration process, microwaves, autoclave, but of these municipalities

IntechOpen

approximately 36% disposed of this waste without any prior treatment, which is not in accordance with current laws, increasing risks to the environment, workers and public health, being sent to landfills, dumps, and septic tanks.

Waste produced by health services can be dangerous, toxic, and lethal because it has a high potential for infection and disease transmission, which can have serious consequences for both the environment and the population. This has already been proven and therefore deserves to be highlighted in discussions due to the impacts it can cause. The high production of BW has caused numerous problems over the years regarding its ideal disposal, raising concerns among the public authorities about implementing measures for its correct disposal, aiming at the possibility of reducing environmental impacts [3, 4].

Biomedical waste requires cold and ventilated storage, high disposal costs for hospitals, strict control of pollutant emissions for municipalities, etc. During the treatment of BW by conventional incineration processes, high levels of polychlorinated dibenzo-p-dioxins and dibenzofurans (which are declared carcinogenic, can pollute the air and landfills, and reduce soil quality) are emitted into the environment by their exhaust gases (77–894 ng/m^3) and residual ash (69–4915 μg/m^3) [3, 5]. The World Health Organization (WHO) [6] recommends on-site high-temperature treatment prior to disposal of this waste to reduce the risk of infection during transport and storage, which brings extra risks of infection and treatment costs. On the other hand, BW physically contains up to 54% paper, 20% textile, 26% organic, 50% plastic, 10% metal, and 15% glass-containing waste, while it consists of up to 35% carbon, 15% hydrogen, 16% nitrogen, 26% oxygen, 1% sulfur, and 3% chlorine, with a maximum calorific value of 8820 kcal/kg [5]. Therefore, the characteristics of BW have many similarities with municipal solid waste (MSW), which can be used for energy production. In addition, this waste can be used as fuel if more environmentally friendly, healthy, and sustainable disposal options are developed [3–5].

There are very few studies on the composition of Brazilian biomedical waste, as this waste is heterogeneous and pathogenic in nature, which inhibits the development of future research. The average composition of Brazil's BW is presented in **Table 1** [3, 7].

Cafure and Patriarcha-Gracioli [8] state that the environmental impacts caused by inadequate management of BW can contaminate and increase hospital infection rates or generate epidemics due to contamination of the water table. Therefore, it is

Components	Average (%)
C	31.6
H	3.74
S	0.20
Cl	0.15
N	0.51
O	37.3
H_2O	20.8
Cinzas	7.0

Table 1.
Average elemental composition of Brazilian BW [7].

essential that technologies aimed at the correct and effective disposal of BW are developed and applied by all Brazilian municipalities and throughout the world.

1.2 Plasma gasification

Plasma gasification technology is recommended for processing municipal, commercial, industrial, petrochemical, and healthcare solid waste. Syngas from plasma gasification are converted into a wide variety of energy products, including electricity, through gas turbines, reciprocating engines and fuel cells, heat and steam [9, 10].

Plasma gasification of waste exploits the thermochemical properties of plasma. The plasma provides the energy needed to maintain the temperature inside the reactor at values sufficient for the dissociation of gas molecules produced by the decomposition of the waste. Due to the high temperature, the inorganic components of the treated materials are melted, the organic components are volatilized, and complex molecules are dissociated. The molten inorganics are removed from the reactor and, upon cooling and solidification, produce a slag-like substance. Organic materials, mainly containing chemically bonded carbon, hydrogen, and oxygen, are decomposed into syngas, which can be used as a high-quality fuel or in the chemical industry [7, 9, 10].

A plasma gasification system consists of the following components: feed system, plasma torches and power supplies, gasification reactor, slag handling equipment, gas treatment system, and monitoring and control system. The feeding system accommodates different types of raw materials and, if necessary, prepares the raw material by grinding before entering the reactor [10, 11].

Plasma gasifiers come in a variety of configurations but typically consist of a reactor connected to a plasma torch that provides the energy required for the gasification reactions. The reactor chamber is made of refractory material that is capable of withstanding the high temperatures of the plasma torch. When the feedstock enters the reactor, the high-temperature plasma decomposes it and the inorganic/ash portion is reduced to a molten slag [3, 12]. And Sikarwar et al. [13] state that due to the high temperature, the gasification of the raw material occurs in milliseconds.

According to Zhao [14], the slag is collected at the bottom of the reactor, where it is removed continuously or periodically. It can be poured into molds to create ingots or bricks or quenched in water to create granules to be reused or disposed of in landfills.

Plasma gasification results essentially in the form of syngas (CO, H_2, and CH_4) and tiny amounts of polluting gases. In this way, the synthesis gas is subjected to cleaning and heat recovery. Once cleaned, it can be burned in the steam or gas cycle or in the gas engine or used in the production of chemicals. The entire process must be monitored and controlled to ensure optimal performance of system temperature, feed rates, plasma power, and slag removal [15].

According to Minutillo [16], the synthesis gas is cooled and cleaned before being used as fuel in an internal combustion engine (ICE) cycle or gas turbine (GT) set. Thus, a cooling and cleaning system must be added to receive the syngas produced in the plasma gasifier in both the combined cycle with ICE and GT.

Plasma gasification has gained prominence due to its syngas production and electricity generation potential in recent years as costs have entered a commercially competitive range. A plasma gasification plant in Utashinai, Japan, has been in operation since 2002 and, as of 2014, gasifies 268 tons of municipal solid waste per day and produces 7.9 MWh of electricity [13].

Plasma gasification technology can also be used to process wastewater. In 2002, Westinghouse Plasma built a plasma gasification plant that serves the cities of Mihama and Mikata (Japan), treating 20 t/day of MSW and 4 t/day of sewage sludge. The syngas is converted to heat for drying the sewage sludge prior to gasification [17].

The main limitations are the high cost of construction, maintenance, and operation due to the high consumption of electricity to generate plasma, resulting in low overall efficiency [13].

According to Munir et al. [18], the operational parameters that affect plasma gasification are the gasifier reaction temperature (approximately 1650–11,600°C), residence time (less than 30 min to 3 h), plasma gas flow rates, oxidant and vapor streams. A shorter residence time is better for achieving a simpler and more stable syngas composition. The source of the plasma gas is critical to the composition of the syngas. For example, N_2 plasma produces more H_2 gas, while steam plasma produces CO and H_2. An oxidizing agent is required for further conversion of carbon into syngas. And the addition of steam in plasma gasification increases the calorific value of syngas from 9.3 to 10.2 MJ/m^3, syngas yield, and fixed carbon conversion [19].

According to Ducharme and Themelis [20], for waste treatment, plasma is preferably generated by direct current (DC) electrical discharge, and two types of devices can be used: transferred and non-transferred arcs. In non-transferred mode, the arc is ignited between two water-cooled electrodes inside the torch, and the plasma exits the torch through a nozzle, which can be one of the electrodes, usually the anode. For transferred arcs, the material to be treated is in direct contact with the auxiliary electrode. This has the advantage of high heat transfer to the treated material and low power loss to the torch body. Some plasma torch systems can be switched between non-transferred and transferred arc mode. However, the use of transferred arcs relies on an electrically conductive auxiliary electrode, for example, a molten metal or slag bath [17].

2. Case study

In order to carry out the case study considering the technical and economic aspects of biomedical waste plasma gasification, the city of São Paulo, located in the state of São Paulo, Brazil, was chosen because it is the largest city in the country, with 11.45 million inhabitants [21], containing 27 thousand establishments registered for biomedical waste (BW) collection [22]. According to data presented by the city of São Paulo [22], in 2022, 39,896 tons of BW were collected from large generators of health waste, while the largest amount was collected in 2023–2039,152 tons. As of August 2024, 26.36 tons of BW have already been collected, with an estimate of reaching more than 39 thousand tons by the end of the year. **Tables 2** and **3** show the production of BW in tons for small and large generators in the city of São Paulo in tons, respectively.

According to the website of the City of São Paulo [22], establishments must make solid waste from the health service available on the days and times determined by the city hall as determined by the Legislation Collegiate Resolution RDC 306 of 12/07/2004 and the BW of groups A, B and E are collected in accordance with CONAMA Resolution 358/2005. In order for waste to be transported, establishments must generate documentation specifying the identification of the sender, the classification, quantity, and type of packaging to which the waste is subjected, and the identification of the transporter and the treatment facility. After collection, the BW is sent to

Year	January	February	March	April	May	June	July	August	September	October	November	December	Total
2016	530	564	44	639	617	652	629	665	615	685	608	625	7471
2017	555	587	667	584	746	665	675	737	699	725	701	667	8008
2018	643	617	730	711	717	725	694	800	697	802	723	711	8569
2019	697	719	757	780	838	705	811	831	748	800	739	702	9128
2020	729	691	734	540	597	710	823	803	824	861	806	833	8951
2021	753	813	895	811	872	890	896	926	889	867	858	858	10,330
2022	791	816	912	822	886	870	861	917	846	838	855	822	10,235
2023	760	731	920	774	898	845	869	913	839	851	843	807	10,053
2024	780	800	841	966	906	891	917	918					7018

Table 2.
BW production (tons)—small generators [22].

Year	January	February	March	April	May	June	July	August	September	October	November	December	Total
2016	2561	2603	2965	2885	2864	2880	2883	2932	2738	2778	2760	2962	33,542
2017	2691	2586	2959	2730	3003	2897	2903	2950	2797	2865	2755	2667	33,803
2018	2723	2541	2956	2869	2912	2801	2806	2931	2718	2928	2728	2650	33,562
2019	2722	2648	2792	2886	2973	2758	2861	2861	2751	2907	2710	2649	33,517
2020	2711	2589	2727	2488	3017	3292	3476	3395	3251	3308	3281	3337	36,872
2021	3370	3334	4188	3869	3921	3857	3731	3635	3405	3426	3280	3279	43,294
2022	3429	3228	3401	3187	3413	3393	3411	3408	3255	3348	3283	3141	39,896
2023	3167	2984	3501	3186	3415	3254	3247	3394	3175	3426	3370	3033	39,152
2024	3099	3067	3391	3434	3463	3222	3342	3344					26,363

Table 3.
BW production (tons)—large generators [22].

Groups	BW production [kg/day]
EGRS I	until 5
EGRS II	5–10
EGRS III	10–20

Table 4.
Classification of small BW generators [23].

treatment units duly licensed by the environmental agency. Once the BW arrives at these units, it is subjected to treatment in autoclaves, where it is exposed to heat (temperatures of up to 150°C) and high-pressure humidity for an appropriate period of time to eliminate all possible microorganisms that could contaminate the environment and harm the health of the population. Autoclave technology ensures levels of microbial inactivation that are established by Brazilian legislation and internationally recognized. After processing in the autoclave, the decontaminated BW undergoes the shredding process and is sent to a landfill.

The city of São Paulo classifies the Establishment Generating Solid Waste from Health Services (EGRS) as small or large generators of BW. Establishments that are considered small generators of BW are divided into three groups, according to **Table 4**. And establishments that are considered large generators are divided into six groups, according to **Table 5**.

The city of São Paulo has instituted the Health Services Solid Waste Tax (TBW) to cover the costs of divisible services of collection, transportation, treatment, and final disposal of health services solid waste, which are mandatory and provided under a public regime within its territory. For each EGRS range, a TBW value is provided; **Tables 6** and 7 present values for small and large generators of BW, respectively. The base year of the rate is 2023, considering the conversion from US$1.00 to R$6.02 [24].

Groups	BW production [kg/day]
EGRS 1	20–50
EGRS 2	50–160
EGRS 3	160–300
EGRS 4	300–650
EGRS 5	650–800
EGRS 6	Above 800

Table 5.
Classification of large BW generators [23].

Groups	quarterly amount [US$]
EGRS especial—I	34.50
EGRS especial—II	46.00
EGRS especial—III	69.01

Table 6.
TBW for BW generators small [23].

Groups	Quarterly amount [US$]
EGRS 1	2197.19
EGRS 2	7030.99
EGRS 3	13183.10
EGRS 4	28563.42
EGRS 5	35154.97
EGRS 6	52733.33

Table 7.
TBW for BW generators large [23].

3. Proposed system

The proposed system consists of a plasma gasifier, a heat exchanger, a cleaning system and an internal combustion engine. The plasma gasifier is powered by BW, the plasma torch transforms it into ashes and syngas (CO and H_2), which is cooled in a heat exchanger and taken to the cleaning system to remove impurities and finally burned in the internal combustion engine to generate electricity. The electrical energy generated can be sold to the electricity distribution network (EDN) and therefore the EDN provides electrical power to the plasma gasifier or directly used in the plasma torches. The flowchart of the plasma gasification system associated with the internal combustion engine is shown in **Figure 1**.

4. Energetic analysis

4.1 Plasma gasifier

The mass conservation equation in the plasma gasifier determines the quantities of syngas produced according to the gasified BW. Thus, the input flows are BW and oxidizing/plasma-forming gas, which are necessary for stabilizing the electric arc, and the output flows are produced gases (syngas + other gases) and ash. The conservation of mass for the plasma gasifier is shown in Eq. (1):

$$\dot{m}_{BW} + \dot{m}_{oxidant} = \dot{m}_{gas} + \dot{m}_{ashes} \tag{1}$$

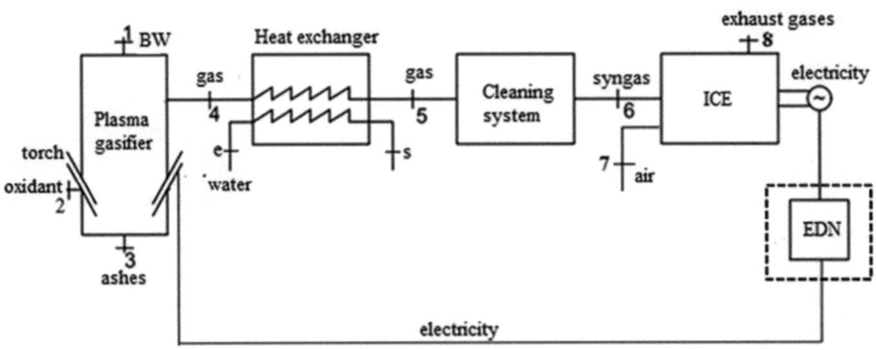

Figure 1.
Plasma gasifier system associated with the internal combustion engine.

where
$\dot{m}_{ashes} = \dot{m}_3$: ashes mass flow [kg/s]
$\dot{m}_{BW} = \dot{m}_1$: BW mass flow [kg/s]
$\dot{m}_{gas} = \dot{m}_4$: gas mass flow [kg/s]
$\dot{m}_{oxidant} = \dot{m}_2$: oxidant mass flow [kg/s]
To determine the amount of syngas produced, Eq. (2) is applied:

$$\dot{m}_{gas} = \dot{m}_{syngas} + \dot{m}_{other.gases} \tag{2}$$

where
$\dot{m}_{gas} = \dot{m}_4$: gas mass flow [kg/s]
$\dot{m}_{other.gases}$: mass flow of other gases that make up the gas [kg/s]
$\dot{m}_{syngas}$: syngas mass flow [kg/s]
The determination of the electricity required in the plasma torch to process the BW is carried out by Eq. (3):

$$E_{torch} = \frac{\dot{m}_{BW}.\Delta H}{\eta_{torch}} \tag{3}$$

where
E_{torch}: electrical power of the torch [kW]
$\dot{m}_{BW}$: biomedical waste mass flow [kg/s]
ΔH: enthalpy difference [kJ/kg], that is, the energy needed to heat 1 kg of BW from the initial state to the end state.
η_{torch} : torch energy efficiency [0,95 or 95%] [10]
The lower heat value of the BW is determined by Eq. (4), considering the mass fractions of each element presented in Section 1.1, **Table 1** [25]:

$$LHV_{BW} = 81\left(X_c - 3\frac{X_o}{8}\right) + \frac{171.X_o}{8} + 345\left(X_H - \frac{X_O}{10}\right) + 25.X_S - 6.(9.X_H + X_{H_2O}) \tag{4}$$

where
LHV_{BW}: syngas lower heating value [kJ/kg]
X_C : carbon mass fraction [−]
X_H : hydrogen mass fraction [−]
X_{H_2O} : water mass fraction [−]
X_N : nitrogen mass fraction [−]
X_O : oxygen mass fraction [−]
X_S : sulfur mass fraction [−]
The energy efficiency of the gasifier is based on the ratio of the energy obtained, syngas, to the energy supplied, BW energy and electricity, according to Eq. (5):

$$\eta_{pg} = \frac{\dot{m}_{syngas}.LHV_{syngas}}{\dot{m}_{BW}.LHV_{BW} + \dot{E}_{torch}} \tag{5}$$

where
E_{torch} :electrical power of the torch [kW]
$\dot{m}_{BW} = \dot{m}_1$: BW mass flow [kg/s]
$\dot{m}_{syngas} = \dot{m}_4$: syngas mass flow [kg/s]

LHV_{BW}: BW lower heating value [kJ/kg]
LHV_{syngas}: syngas lower heating value [kJ/kg]
η_{pg}: plasma gasifier energy efficiency [−]

4.2 Heat exchanger

The syngas leave the plasma gasifier at a high temperature, so it needs to be cooled so that it can pass through the cleaning system and then be burned in an internal combustion engine. According to Xavier [26], the gas inlet temperature in the internal combustion engine is approximately 733 K. Eq. (6) shows the heat power that can be removed from the gas to another fluid, so that the gas leaves the heat exchanger at 733 K:

$$Q_{syngas} = \dot{m}_4.c_{pm.gas}.(T_4 - T_5) \quad (6)$$

where
$c_{pm.gas}$: average specific heat at constant pressure of gas [kJ/kg.K]
$\dot{m}_4$: mass flow at point 4 of the plasma gasifier [kg/s]
Q_{syngas}: heat flux produced in the heat exchanger by the syngas [kW]
T_4: gas temperature at point 4 [K]
T_5: gas temperature at point 5 [K]

4.3 Cleaning system

In order for the gas to reach the internal combustion engine with the appropriate quality, it must pass through a cleaning system. The cleaning system usually involves the removal of elements composed of sulfur, chlorides, mercury, other volatile metals, and particulate matter.

According to Villela [27], the pressure loss in the heat exchanger is 5%, and the same consideration is adopted for the cleaning system; thus, the inlet pressure (P5) is determined by Eq. (7):

$$P_5 = \frac{P_6}{1 - \Delta P_{CL}} \quad (7)$$

where
P_5: syngas pressure at point 5 [kPa]
P_6: syngas pressure at point 6 [kPa]
ΔP_{CL} : pressure drop in the cleaning system [−]

4.4 Internal combustion engine

Commercially available internal combustion engines are powered by natural gas or biogas. When another type of biofuel, such as synthesis gas, which has a lower heat value (LHV), is used, the final performance of the engine changes, mainly in relation to electricity production and electrical efficiency. Thus, to determine the electricity generation efficiency for an engine that uses another type of fuel, it can be determined by Eq. (8) which uses a correction factor [28–29]:

$$\gamma_{LHV} \cdot \eta_{el.ICE} = \frac{\dot{E}_{p.ICE}}{\dot{E}_{syngas.}} \quad (8)$$

where
$\dot{E}_{syngas}$: syngas energy supplied [kW]
$\dot{E}_{p.ICE}$: ICE electric power [kW]
$\eta_{el.ICE}$: ICE electricity generation efficiency [kW]
γ_{LHV}: LHV correction factor [−]
The energy supplied by the syngas is given by Eq. (9):

$$\dot{E}_{syngas} = \dot{m}_{gas}.LHV_{syngas} \quad (9)$$

where
$\dot{E}_{syngas}$: syngas energy supplied [kW]
$\dot{m}_{syngas}$: syngas mass flow[kg/s]
LHV_{sygas}: syngas lower heating value [kJ/kg]
The lower heat value correction factor is calculated according to Eq. (10) [28]:

$$\gamma_{LHV} = 0{,}102.\frac{LHV_{syngas}}{LHV_{NG}} + 0{,}897 \quad (10)$$

where
LHV_{sygas}: syngas lower heating value [kJ/kg]
LHV_{NG}: natural gas lower heating value [kJ/kg]
γ_{LHV}: LHV correction factor [−]
The syngas lower calorific value is determined from the respective syngas composition, according to Eq. (11) [3, 10]:

$$LHV_{syngas} = \sum X_i LHV_i \quad (11)$$

where
LHV_{syngas}: gas lower heating value [kJ/kg]
LHV_i : lower heating value of ith component [kJ/kg]
X_i: mass fraction of the ith component [−]
Table 8 shows the composition (w.t.) of the gas resulting from the BW plasma gasification of Brazil, according to the composition presented in Section 1.1, **Table 1**. After passing through the cleaning system, the syngas have the composition of H_2, CO, and CH_4.

The LHVi is the lower calorific value of the i-th component of the mixture (H_2, CO, and CH_4). The LHV of H_2, CO, and CH_4 is 120,000, 11,084, and 50,023 kJ/kg, respectively [26]. According to Maya et al. [30], the electricity generation efficiency of the internal combustion engine ranges from 14 to 26%. Therefore, for this work, the electricity generation efficiency of the ICE is considered to be 20%.

Elements	H_2	H_2O	HCl	H_2S	N_2	CO	CO_2	CH_4
Mass fraction [−]	0,0535	0,0829	0,0016	0,0022	0,0054	0,6365	0,2121	0,0057

Table 8.
Composition of the elements of the gas [7–10].

4.5 Plasma gasification system associated with the internal combustion engine

The percentage of electricity produced by burning gas in the internal combustion engine that can supply the plasma gasifier is determined by Eq. (12) [7–10]:

$$\Omega_{el.ICE} = \frac{\dot{E}_{p.ICE}}{\dot{E}_{torch}}.100 \tag{12}$$

where

$\dot{E}_{p.ICE}$: ICE electric power [kW]

E_{torch} :electrical power of the torch [kW]

$\Omega_{el.ICE}$: Percentage of electricity that the ICE can supply in the plasma gasifier [%]

The energy efficiency of the plasma gasification process associated with the internal combustion engine is determined using Eq. (13):

$$\eta_{pg.el.ICE} = \eta_{gp}.\eta_{el.MCI} \tag{13}$$

5. Economic analysis

To assess the viability of the gasification process in terms of profitability, it is necessary to conduct economic engineering studies. In order to obtain financing to invest in this new technology, the annuity factor must be considered, which is a function of the annual interest rate and the amortization period (payback). The annuity factor is calculated using Eq. (14) [10, 29]:

$$f = \frac{q^k(q-1)}{q^k - 1} \tag{14}$$

where

$$q = 1 + \frac{r}{100} \tag{15}$$

where

f : annuity factor [1/year]

k: payback [years]

r: annual interest rate [%]

5.1 Syngas production cost

The determination of the syngas production cost, considering plasma gasification associated with the internal combustion engine, is carried out using the methodology adapted from Paulino et al. [29], according to Eq. (16):

$$C_{syngas} = \frac{Inv_{pg.plant}.f}{H.\dot{E}_{syngas}} + Cop_{pg.plant} + Cman_{pg.plant} \tag{16}$$

where

C_{syngas}: syngas production cost [US$/kWh]

$Cman_{gp.plant}$: plasma gasification plant maintenance cost [US$/kWh]
$Cop_{gp.plant}$: plasma gasification plant operation cost [US$/kWh]
$\dot{E}_{p.ICE}$: ICE electrical power [kW]
$\dot{E}_{syngas}$: energy supplied by the syngas [kW]
$\dot{E}_{torch}$: plasma torch power [kW]
f : annuity factor [1/year]
H: equivalent period of utilization [h/year]
$Inv_{gp.plant}$: plasma gasification plant investment [U$S]

To determine the operating cost for each configuration, the electricity required by the torch that needs to be purchased from the power distribution network is considered, as per Eq. (17):

$$Cop_{pg.plant} = \frac{E_{torch} * P_{el.}}{\dot{E}_{syngas}} \tag{17}$$

where
$Cop_{pg.plant}$: plasma gasification plant operation cost [US$/kWh]
$\dot{E}_{torch}$: plasma torch power [kW]
$\dot{E}_{syngas}$: energy supplied for syngas [kW]
P_{el}: electricity tariff from utility [US$/kWh]

According to Paulino [3], the maintenance cost of the plasma gasification system is 3% of the plant investment portion, as per Eq. (18):

$$Cman_{gp.plant} = 0,03.\frac{Inv_{gp.plant}.f}{H.\dot{E}_{gás}} \tag{18}$$

5.2 Cost of electricity production in the internal combustion engine

The cost of electricity produced in the internal combustion engine is determined by Eq. (19), as adapted by Paulino et al. [3]:

$$C_{el.MCI} = \frac{Inv_{MCI}.f}{H.\dot{E}_{p.MCI}} + Cop_{MCI} + Cman_{MCI} \tag{19}$$

where
$C_{el.ICE}$: electricity production cost at ICE [US$/kWh]
$Cman_{ICE}$: ICE maintenance cost [US$/kWh]
Cop_{ICE}: ICE operation cost [US$/kWh]
$\dot{E}_{p.ICE}$: ICE electrical power [kW]
$\dot{E}_{syngas}$: energy supplied by syngas [kW]
f : annuity factor [1/year]
H: utilization equivalent period [h/year]
Inv_{ICE}: ICE investment [U$S]

The operating cost of the internal combustion engine considers the cost of producing the synthesis gas and the electricity it produces and is determined using Eq. (20):

$$Cop_{ICE} = \frac{C_{syngasICE}.\dot{E}_{syngas}}{\dot{E}_{p.ICE}} \tag{20}$$

According to Brizi [31], the maintenance cost of the internal combustion engine is 0.0130 US$/kWh.

5.3 Investments

To set up a plasma gasification plant with gas production and electricity generation, it is necessary to invest in equipment. The equations in this section show the investment in each piece of equipment used in the plants under study.

The investment in the plasma gasification system considers the plasma gasifier, heat exchanger to cool the syngas and produce hot water and the cleaning system, as per Eq. (21):

$$Inv_{pg.plant} = Inv_{pg} + Inv_{HE} + Inv_{CS} \tag{21}$$

where
Inv_{pg} : plasma gasifier investiment [US$]
$Inv_{gp.plant}$: plasma gasification plant investment [U$S]
Inv_{CS}: cleaning system investment [US$]
Inv_{HE}: heat exchanger investment [US$]

The investment cost of a plasma gasification plant using water vapor in the torch is 4800 euros per kg/h multiplied by the factor attributed by Gasser [32], according to Eq. (22):

$$Inv_{gp} = \left(4800 * 1,25 * \dot{m}_{BW.h} * ER_{US\$/EUR}\right) \tag{22}$$

where
$Inv_{gp.plant}$: plasma gasification plant investment [U$S]
$\dot{m}_{BW.h}$: BW mass flow per hour [kg/h]
$ER_{US\$/EUR}$: US$/EUR exchange rate [−]

The investment of a heat exchanger is expressed by eq. (23) [26]:

$$Inv_{HE} = 60 * Q_{syngas} \tag{23}$$

where
Inv_{HE}: heat exchanger investment [US$]
Q_{syngas}: heat flux produced in the heat exchanger by the syngas [kW]

The investment cost of the cleaning system in the range of 3 to 1860 kg/min is determined through Eq. (24):

$$Inv_{CS} = 1250 * \left(\frac{m_{syngas.min}}{928,5}\right)^{0,74} * 10^2 \tag{24}$$

where
Inv_{CS} : cleaning system investment [US$]
$m_{syngas.min}$: syngas mass flow per minute in the cleaning system [kg/min]

Xavier [26] developed Eq. (25) for the investment in the internal combustion engine:

$$Inv_{MCI} = 400 * \left(\frac{\dot{E}_{p.MCI}}{455}\right)^{0,9021} * 10^3 \tag{25}$$

where
$\dot{E}_{p.ICE}$: ICE electrical power [kW]
Inv_{ICE}: ICE investment [U$S]

5.4 Annual saving expected

To assess the return on investment period, the annual saving expected is determined, which considers the sale price of electricity, the cost and production of electricity for each configuration, the cost of disposing of the BW, the equivalent period of use, and the electricity required by the torch, according to Eq. (26) (adapted Paulino [29]):

$$R_{annual} = \left[\left(P_{s.el.ICE} - C_{el.ICE}\right) * \left(\dot{E}_{p.ICE} * H\right)\right] + C_{disp.waste} - \left[P_{el.} * H * E_{torch}\right] \qquad (26)$$

where
$C_{el.ICE}$: electricity production cost at ICE [US$/kWh]
$C_{disp.waste}$: waste disposal cost [US$/year]
$\dot{E}_{p.ICE}$: ICE electrical power [kW]
H: utilization equivalent period [h/year]
P_{el}: electricity tariff from distribution network [US$/kWh]
$P_{s.el.ICE}$: electricity produced sales price at ICE [US$/kWh]
R_{annual}: annual saving expected [US$/year]

For the economic analysis, the maximum production of each group and the respective Health Services Solid Waste Tax (TBW) per kg will be considered. **Tables 9** and **10** show the TBW in R$/kg for small and large generators, respectively.

Furthermore, for the purpose of making a comparison of the economic analysis, the incineration scenario is also considered. It is considered that instead of processing

Groups	[US$/kg]
EGRS I	0.08
EGRS II	0.05
EGRS III	0.04

Table 9.
TBW for BW generators small [23].

Groups	[US$/kg]
EGRS 1	0.49
EGRS 2	0.49
EGRS 3	0.49
EGRS 4	0.49
EGRS 5	0.49
EGRS 6	under 0.73

Table 10.
TBW for BW generators large [23].

Point	$\dot{m}$ [kg/s]	T [° C]	P [kPa]	h [kJ/kg]	s [kJ/kg.K]	c_p [kJ/kg.K]
1	1,24	25	100	—	—	—
2	—	—	—	—	—	—
3	0,09	767	100	—	—	—
4	1,15	767	116	—	—	1.877
5	1,15	460	110	—	—	1.869
e	0,15	25	100	105	0,367	4.180
s	0,15	100	100	2676	7361	—
6	1,15	460	105	—	—	1.869
7	2,69	25	100	—	—	1.004
8	3,84	377	100	—	—	1.118

Table 11.
Thermodynamic properties.

Properties	symbol	value
Average specific heat at constant pressure of gas [kJ/kg.K]	$c_{pm.gas}$	1,87
Syngas energy supplied [kW]	$E_{syn.gas}$	15,861
BW lower heating value [kJ/kg]	LHV_{BW}	19,006
Syngas lower heating value [kJ/kg]	LHV_{syngas}	13,754
Heat flux produced in the heat exchanger by the gas – ICE [kW]	Q_{syngas}	663
Plasma gasifier energy efficiency [−]	η_{pg}	49,33
Energy efficiency of the plasma gasification process associated with the ICE [−]	$\eta_{pg.el.ICE}$	9,87
Enthalpy difference [kJ/kg]	ΔH	5673

Table 12.
Energetic analysis results.

the BW in an autoclave, shredding it, and then taking it to the landfill, the BW is incinerated. And according to LUFTECH Environmental Solutions (2021), the average amount charged per kg of hospital waste is US$0.41 to US$0.83 (depending on the competition). For this work, the average value of US$0.62 is considered.

6. Results and discussions

6.1 Energy analysis results

The results of the energy analysis were obtained considering the production of 39,152 tons of biomedical waste in the year 2023 in the city of São Paulo – Brazil, only the large generators. For the BW plasma gasification process considering the composition of Brazil, Paulino [10] found that there was no need to add an oxidizing agent, therefore, there will be no mass flow of oxidant at point 2. Thus, **Tables 11** and **12** presents the properties of each point of the plasma gasification system associated with the internal combustion engine, as presented in Section 3, **Figure 2**.

The syngas produced in the BW plasma gasification can generate 2938.43 kW of electrical power in an internal combustion engine. Therefore, the Caterpillar DM5398 (3105 kW) engine was selected. And **Figure 3** shows energetic analysis results.

Figure 3 shows the electrical power required in the torch, the potential for electricity production in the internal combustion engine using syngas from plasma

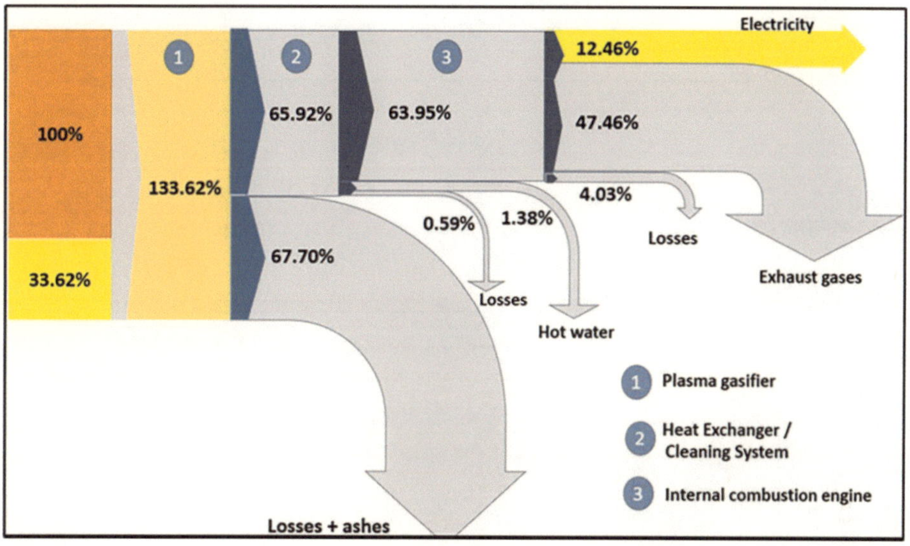

Figure 2.
Sankey diagram.

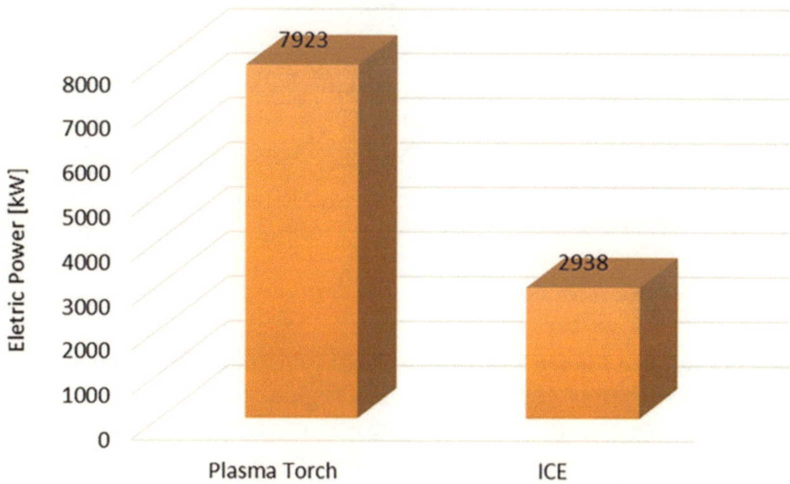

Figure 3.
Electric power (plasma torch and ICE).

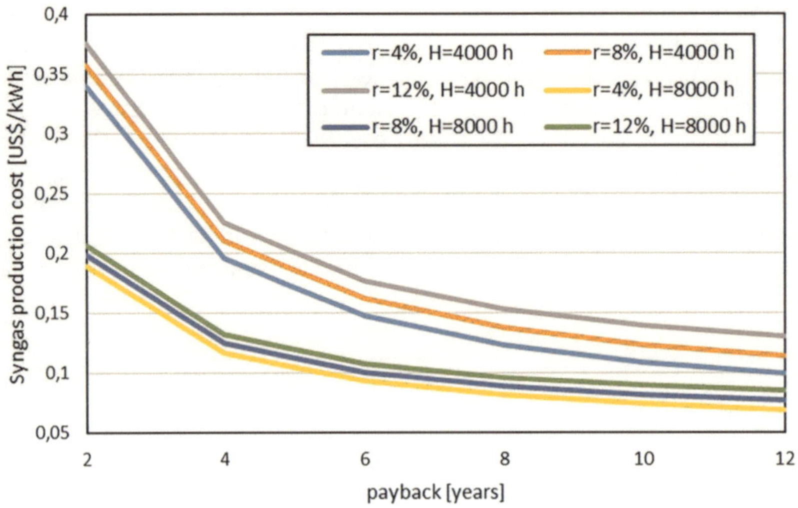

Figure 4.
Syngas production cost.

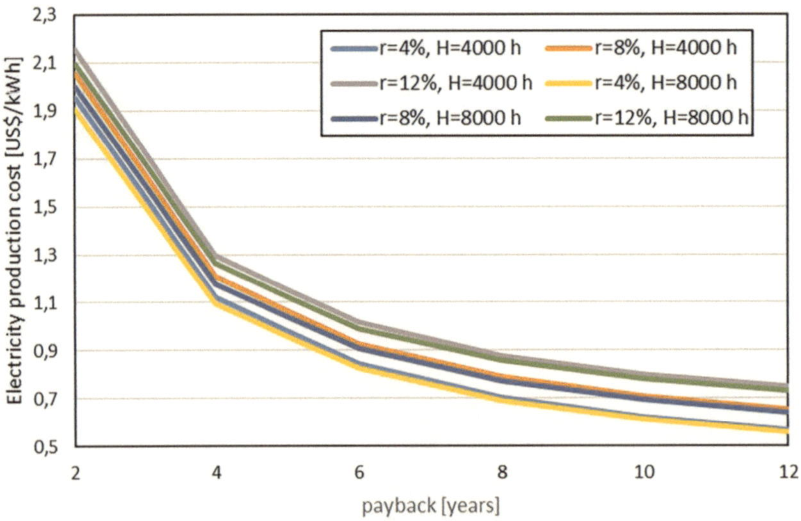

Figure 5.
Electricity production cost.

gasification of BW, and **Figure 2** shows the Sankey diagram of the plasma gasification system associated with the internal combustion engine.

It can be seen from **Figure 3** that the electricity demand required by the torch is greater than that produced by the internal combustion engine. The internal combustion engine supplies only 37.1% of what the torch requires for the operation of the plasma gasification system, and this is due to losses in the gasifier and the low efficiency of the engine using syngas, as shown in **Figure 2**.

6.2 Results of economic analysis

The economic analysis was carried out considering the following parameters:

- Electricity rate for the city of São Paulo is 0.106 US$/kWh – low voltage B3 [33].
- Sale price of electricity from waste in Brazil is 0.100 US$/kWh [34].
- Annual interest rates of 4%, 8%, and 12%.
- Equivalent period of system use is 4000 and 8000 h/year.
- Amortization period of invested capital, payback, ranging from 2 to 12 years.
- Dollar exchange rate in relation to the Euro, US$1.00 = EUR 0.95 [24].
- Real exchange rate in relation to the Dollar, US$1.00 = R$6.02 [24].

Figures 4 and **5** show the syngas and electricity production cost as a function of the payback for the plasma gasification system associated with the internal combustion engine.

Through **Figure 4**, it is possible to observe that the cost of syngas production decreases with an increase in payback and increases with an increase in the annual interest rate. The syngas production cost varies from 0.0.09 to 0.37 US$/kWh and 0.0.06–0.21 US$/kWh for the usage period of 4000 and 8000 h/year, respectively. It can be seen from **Figure 5** that the cost of electricity production for the period of use of 4000 and 8000 h/year are very similar due to the same equipment investment value, and in both situations, they are higher than the electricity tariff in the city of São Paulo (0.106 US$/kWh), due to the high cost of syngas used in the internal

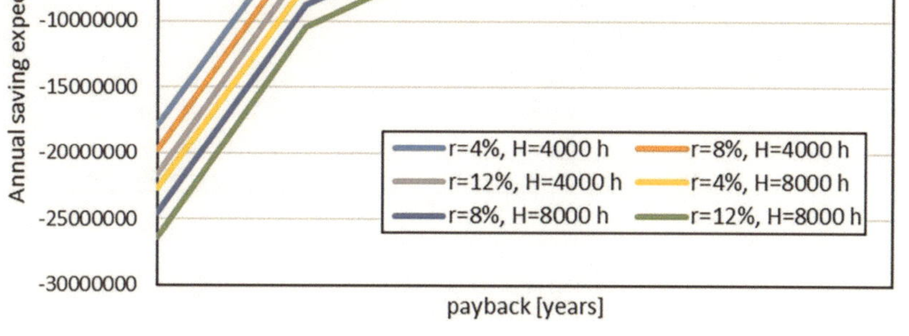

Figure 6.
Annual saving expected, subtracting autoclave.

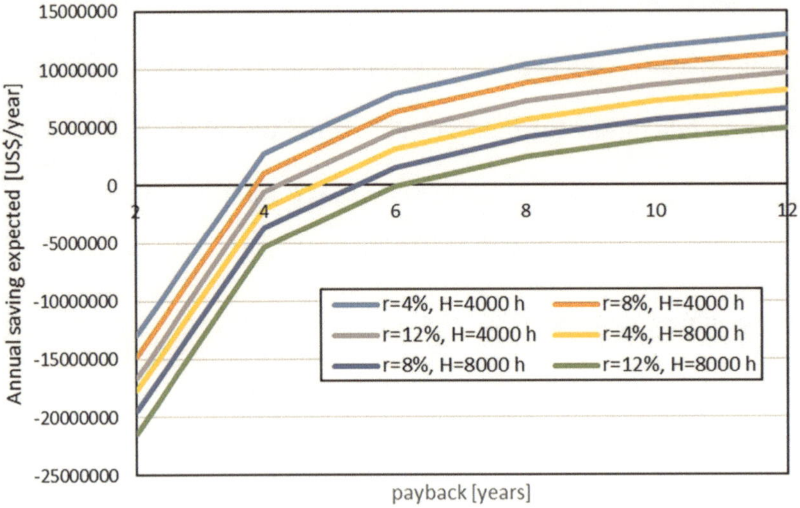

Figure 7.
Annual saving expected, subtracting incineration.

combustion engine. The cost of electricity production varies from 0.56 to 2.15 US$/kWh and 0.55 to 2.09 US$/kWh for the period of use of 4000 and 8000 h/year.

To calculate the annual savings expected, the plasma gasification process for the disposal of BW is considered as the cost that would arise if the autoclaving process used in the city of São Paulo were carried out and also the incineration process, which is the most used in Brazil, is subtracted. **Figures 6** and **7** show the annual revenue as a function of the payback, considering the subtraction of the cost of disposing of BW for autoclaving and incineration, respectively. And to compare the processes, **Figure 8** was prepared for the period of use of 4000 h/year.

Figures 6 and **7** show that the payback period is higher for the 8000 h/year usage period due to the high cost of electricity used for a longer period of time in the plasma

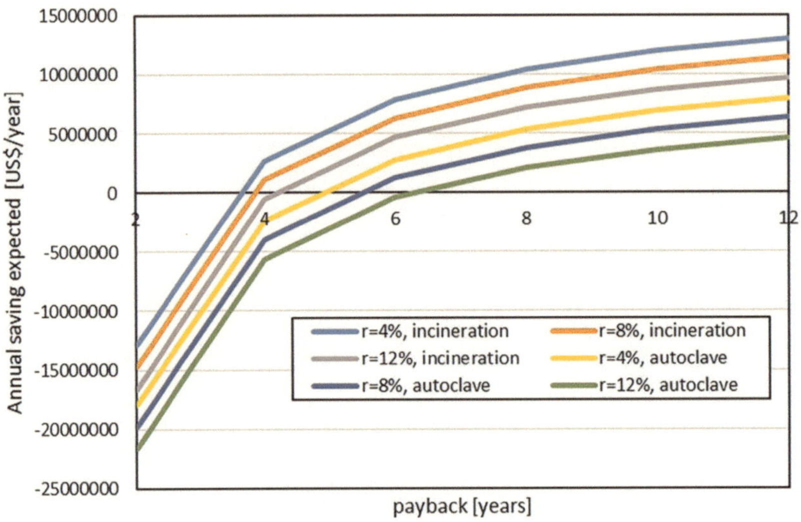

Figure 8.
Annual saving expected, comparison between autoclave and incineration.

gasification process. **Figure 8** shows that for the 4000 h/year usage period, which is the most viable scenario, the payback is between 5 and 8 years and 3 and 5 years for the plasma gasification process, subtracting the cost of the autoclave and incineration processes, respectively. In both cases, the gasification process is attractive due to the payback and the fact that a plant like the one studied typically has a useful life of 30 years. Although the payback period for replacing the autoclave process is higher than for the incineration process, it should be considered that after the autoclave process, the BW will still be sent to landfills and occupy large areas.

7. Conclusions

Plasma gasification technology is still under development and needs to be more widespread so that it can evolve in terms of efficiency and become more economically attractive. Through energy analysis, it was possible to obtain the necessary properties at each point and thus determine that the plasma gasifier requires 7923 kW of electrical energy to process 39,152 tons of BW per year in the city of São Paulo and that the plasma gasifier has an energy efficiency of 49.33%; when associated with the ICE, it has the potential to produce 2938 kW, providing 37.1% of the total electrical power required in the torches. From a technical point of view, gasification technology is not very attractive due to its high electricity consumption and the fact that the internal combustion engine associated with the system does not produce all the necessary energy. Therefore, studies must be carried out to improve the efficiency of the devices and, consequently, the overall efficiency of the system. The economic analysis shows that the cost of syngas production is in the range of 0.06–0.37US$/kWh, with electricity production costs between 0.55 and 2.15 US$/kWh. These values are much higher than the electricity tariff in the city of São Paulo (US$0.106/kWh). However, when analyzing the payback, plasma gasification technology (the useful life of 25–30 years) shows promise, with a payback of 3–8 years considering the use of plasma gasification rather than autoclave and incineration processes. Finally, it is possible to point out that although plasma gasification technology is not technically attractive from the point of view of electrical self-sufficiency, it is promising economically and environmentally correct and a good alternative for processing BW (when compared to the current processes used autoclave and incineration), as it allows energy recovery from the waste and considerably reduces environmental pollution. Thus, plasma gasification becomes an option to replace the incineration and autoclaving processes most used in Brazil, and can also encourage cities that do not collect or dispose of BW in inappropriate places to begin the appropriate treatment of this waste.

Author details

Regina Franciélle Silva Paulino[1*], Alexei Mikhailovich Essiptchouk[2] and José Luz Silveira[3]

1 UniMetrocamp University Center – Wyden and Institute of Bioenergy Research (IPBEN-UNESP)-Associated Laboratory of Guaratinguetá, Campinas, Brazil

2 Sao Paulo State University (UNESP), São José dos Campos, Brazil

3 Laboratory of Optimization of Energy Systems (LOSE), Department of Energy, School of Engineering and Institute of Bioenergy Research (IPBEN-UNESP)-Associated Laboratory of Guaratinguetá, Sao Paulo State University (UNESP), São Paulo, SP, Brazil

*Address all correspondence to: regina.paulino@professores.unimetrocamp.edu.br

IntechOpen

© 2025 The Author(s). Licensee IntechOpen. This chapter is distributed under the terms of the Creative Commons Attribution License (http://creativecommons.org/licenses/by/4.0), which permits unrestricted use, distribution, and reproduction in any medium, provided the original work is properly cited.

References

[1] BRASIL. Lei de n° 12.305: Política Nacional de Resíduos Sólidos (PNRS). Brasília: Presidência da República (Internet); 2024. Available from: http://www.planalto.gov.br/ccivil_03/_Ato2007-2010/2010/Lei/L12305.htm [Accessed: October 15, 2024]

[2] ABREMA. Brazilian Association of Waste and Environment. Panorama of Solid Waste in Brazil 2023 (Internet). 2024. Available from: https://www.abrema.org.br/wp-content/uploads/dlm_uploads/2024/03/Panorama_2023_P1.pdf [Accessed: October 18, 2024]

[3] Paulino RFS, Essiptchoukb AM, Silveira JL. The use of syngas from biomedical waste plasma gasification systems for electricity production in internal combustion: Thermodynamic and economic issues. Energy. 2020;**199**: 117419. DOI: 10.1016/j.energy.2020.117419

[4] Sena RM, Sanchez MCO, Moraes EB, Xavier ML, Braga ALS, Porto MAOP. Health management in Brazil: Challenges of managers and health professionals waste Gestión de residuos sanitarios en Brasil: Retos de gestores y profesionales de la salud. Research, Society and Development. 2021;**10**(4):14510413960. DOI: 10.33448/rsd-v10i4.13960

[5] Erdogan AA, Yilmazoglu MZ. Plasma gasification of the medical waste. International Journal of Hydrogen Energy. 2021;**46**(57):29108-29125. DOI: 10.1016/j.ijhydene.2020.12.069

[6] World Health Organization. Water, Sanitation, Hygiene, and Waste Management for SARS-CoV-2, the Virus that Causes COVID-19: Interim Guidance (2020). 2020. Available from: https://iris.who.int/bitstream/handle/10665/333560/WHO-2019-nCoV-IPC_WASH-2020.4-eng.pdf. p. 29 [Accessed: November 12, 2024]

[7] Paulino RFS, Essiptchouk AM, Costa LPC, Silveira JL. Thermodynamic analysis of biomedical waste plasma gasification. Energy. 2022;**244**:122600. DOI: 10.1016/j.energy.2021.122600

[8] Cafure VA, Patriarcha-gracioli SR. Health service waste and its environmental impacts: A literature review. Campo Grande: Interações. 2015; **16**(2):301-314. DOI: 10.1590/151870122015206

[9] Zhovtyansky V, Valinčius V. Efficiency of plasma gasification technologies for hazardous waste treatment. In: Yun Y, editor. Gasification for Low-Grade Feedstock. Korea, South: Institute for Advanced Engineering; 2018. pp. 142-189. DOI: 10.5772/intechopen.74485

[10] Paulino RFS, Silveira JL. Plasma gasification of biomedical waste: Energetic and exergetic aspect. In: Pedro M, Fernandez FGA, editors. Sustainable Development Goals Series. 1st ed. Universidad Politecnica de Madrid; Chemical Engineering University of Almeria: Madrid, Spain; Almeria, Spain. 2024. pp. 211-238

[11] Janajreh I, ShabbarRaza SS, Valmundsson AS. Plasma gasification process: Modeling, simulation and comparison with conventional air gasification. Energy Conversion and Management. 2013;**65**:801-809. DOI: 10.1016/j.enconman.2012.03.010

[12] Moustakas K, Fatta MS, Haralambous K, Loizidou M. Demonstration plasma gasification/

vitrification system for effective hazardous waste treatment. Journal of Hazardous Materials. 2005;**123**(1–3): 120-126. DOI: 10.1016/j.jhazmat.2005.03.038

[13] Sikarwar VS, Zhao M, Clough P, Yao J, Zhong X, Mohammad MZ, et al. An overview of advances in biomass gasification. Energy & Environmental Science. 2016;**9**:2939-2977. DOI: 10.1039/C6EE00935B

[14] Zhao P, Ni G, Jiang Y, Chen L, Chen M, Meng Y. Destruction of inorganic municipal solid waste incinerator fly ash in a DC arc plasma furnace. Journal of Hazardous Materials. 2010;**181**(1–3):580-585. DOI: 10.1016/j.jhazmat.2010.05.052

[15] Mountouris A, Voutsas E, Tassios D. Solid waste plasma gasification: Equilibrium model development and exergy analysis. Energy Conversion and Management. 2006;**47**(13–14):1723-1737. DOI: 10.1016/j.enconman.2005.10.015

[16] Minutillo M, Perna A, Di Bona D. Modelling and performance analysis of an integrated plasma gasification combined cycle (IPGCC) power plant. Energy Conversion and Management. 2009;**50**(11):2837-2842. DOI: 10.1016/j.enconman.2009.07.002

[17] Hrabovsky M, Van der Walt IJ. Plasma waste destruction. In: Kulacki FA, editor. Handbook of Thermal Science and Engineering. Minneapolis, USA: Springer; 2017. DOI: 10.1007/978-3-319-32003-8_32-1

[18] Munir MT, Mardon I, Al-Zuhair S, Shawabkeh A, Saqib NU. Plasma gasification of municipal solid waste for waste-to-value processing. Renewable and Sustainable Energy Reviews. 2019;**116**:109461. DOI: 10.1016/j.energy.2020.117419

[19] Tendler M, Rutberg P, Oost GV. Plasma based waste treatment and energy production. Plasma Physics and Controlled Fusion. 2005;**47**:219-230. DOI: 10.1088/0741-3335/47/5A/016

[20] Ducharme C, Themelis N. Analysis of thermal plasma—Assisted waste-to energy processes. In: Proceedings of the 18th Annual North American Waste-to-Energy Conference NAWTEC18. New York, NY: Columbia University; 2010

[21] IBGE. Brazilian Institute of Geography and Statistics. Available from: https://www.ibge.gov.br/cidades-e-estados/sp/sao-paulo.html [Accessed: December 13, 2024]

[22] São Paulo City Hall. Available from: https://capital.sp.gov.br/web/spregula/w/residuos_solidos/menu/185375 [Accessed: December 13, 2024]

[23] Municipal Secretary of Finance of São Paulo. Available from: https://capital.sp.gov.br/web/fazenda/w/servicos/taxaderesiduos/2356 [Accessed: December 13, 2024]

[24] UOL Economy. Available from: https://economia.uol.com.br/cotacoes/cambio/ [Accessed: December 13, 2024]

[25] El-Salam MM. Hospital waste management in El-Beheira governorate. Egypt Journal of Environmental Management; 2010;**91**(3):618-629. DOI: 10.1016/j.jenvman.2009.08.012

[26] Xavier BH. Thermodynamic, ecological and economic aspects of cogeneration systems with internal combustion engines operating on natural gas, biogas and synthesis gas [Masters dissertation]. Portuguese: School of Engineering, Campus Guaratinguetá, UNESP; 2016

[27] Villela ISC. Development of a thermoeconomic model that considers environmental impacts [thesis]. Portuguese: School of Engineering, Campus Guaratinguetá, UNESP; 2007

[28] Wang J, Mao T, Sui J, Jin H. Modeling and performance analysis of CCHP (combined cooling, heating and power) system based on co-firing of natural gas and biomass gasification gas. Energy. 2015;**93**:801-815

[29] Paulino RFS, Silveira JL. Biohydrogen production and bagasse gasification process in the sugarcane industry. Sustainable Bioenergy: Advances and Impacts; 2015;**2E**:89-126. DOI: 10.1016/B978-0-12-817654-2.00004-6

[30] Maya DMY, Sarmiento ALE, Oliveira CAB d S, Lora EES, Andrade RVJ. Gasification of municipal solid waste for power generation in Brazil, a review of available technologies and their environmental benefits. Chemical Engineering Journal. 2016;**10**: 249-255. DOI: 10.17265/1934-7375/2016.06.001

[31] Brizi F, Silveira JL, Desideri U, Reis JA, Tuna CE, Lamas WQ. Energetic and economic analysis of a Brazilian compact cogeneration system: Comparison between natural gas and biogas. Renewable and Sustainable Energy Reviews. 2014;**38**:193-211. DOI: 10.1016/j.rser.2014.05.088

[32] Gasser M. Feasibility Study on the Construction and Maintenance of a Middle Scale Plasma Gasification Plant in Effurun, Nigeria. IAESTE-internship report International; 2019; 29p. Available from: https://www.researchgate.net/publication/330654633_Feasibility_Study_on_the_Construction_and_Maintenance_of_a_middle_scale_Plasma_Gasification_Plant_in_Effurun_Nigeria [Accessed: November 10, 2019]

[33] Enel. Available from: https://www.enel.com.br/pt-saopaulo/Para_Voce/tarifa-energia-eletrica.html [Accessed: December 15, 2024]

[34] Canal Energy. Available from: https://www.canalenergia.com.br/noticias/53221995/precos-do-a-5-variam-de-r-19496mwh-a-mais-r-600mwh [Accessed: December 13, 2024]